"AN EMINENT LIVERPOOL BOTANIST"

A LIFE OF JAMES ALFRED WHELDON

MSC, ALS, ISM

(1862-1924)

"AN EMINENT LIVERPOOL BOTANIST"

A LIFE OF JAMES ALFRED WHELDON

MSC, ALS, ISM

(1862-1924)

WILLIAM PETER WHELDON

Published by B Wheldon

A CIP catalogue record for this book is available from the British Library.

ISBN 978-0-9570080-0-7

Cover design by Clare Brayshaw

Prepared and printed by:

York Publishing Services Ltd
64 Hallfield Road
Layerthorpe
York YO31 7ZQ

Tel: 01904 431213

Website: www.yps-publishing.co.uk

CONTENTS

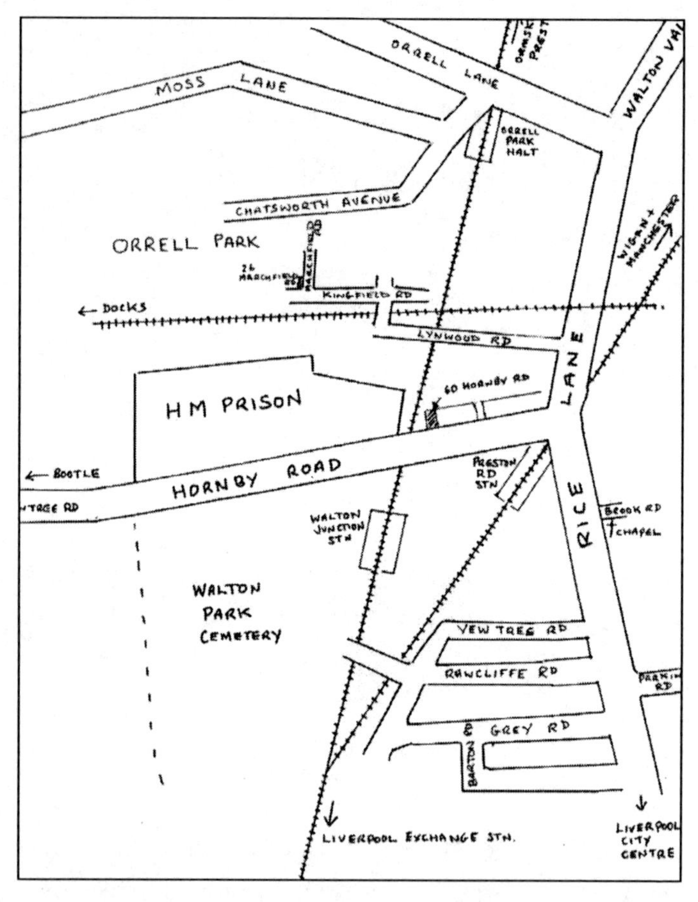

A sketch map of the Walton area, showing some of the places mentioned in the text. It is not to scale and some roads and streets have been left out for clarity. Not all the features were present at any one time. On the extreme edges of the map are Aintree Road, leading towards Bootle which was not completed during Wheldon's lifetime, and Parkinson Road, home of Septimus Simpson. The railway line beyond Orrell Park goes to Ormskirk and Preston, and Preston Road station is now called Rice Lane on the Merseyrail line from Liverpool Central to Kirkby. Rice Lane is the current A59.

FOREWORD

The following is an extract from the diary of Jim Wheldon, grandson of James Alfred Wheldon.

Wednesday 25th July 1979

"Left on the 9.15, arrived at Walton Junction about 10.30 and found my grandfather's grave in Walton Park Cemetery. Absolutely overgrown by brambles, in a wilderness that was once a well-kept, green-verged, shale-dressed, upper-class resting place. It is owned by St Nicholas's, Liverpool. Perhaps I should write to the Rector, suggesting preservation as a nature reserve. Saw 26 Marchfield Road and 64 Stanley Gardens where 88-year-old Mrs Fielding was amazed that little Jimmy had come back!

So much demolished – Brook Road Church, Rice Lane School. Perhaps A.E. Housman was right. Perhaps it is better not to shatter memories."*

I was with my father on that day. He had just retired from teaching and I had a day off from my job with British Rail. He was always very interested in the life and work of his grandfather, and had been named after him, and at the start of his retirement was keen to revisit the grave and locations in Walton where he had grown up and where his grandfather had lived and worked. And he was keen to pass on that interest to me, which he did.

I distinctly remember Mrs Fielding, cutting her hedge and recognising him almost instantly although he had left Stanley Gardens as a ten-year-old in 1930. Although he never wrote to the Rector, he was remarkably prophetic and today the cemetery is part of Rice Lane City Farm, the brambles have been cut back and the grave remains in very good condition in well-kept grounds.

I had always been aware of James Alfred Wheldon and of the *Flora of West Lancashire* on the bookshelf, and of the Imperial Service Medal in the cupboard and I remember my grandmother, Madge, talking about him quite frequently, all of which increased my curiosity.

This work is dedicated to my father Jim Wheldon (1920-2009). He collected and preserved some of the material upon which it is based and passed on to me a life-long interest in the family history and times past in general.

It is not intended to be a botanical textbook. It is intended to be about the man himself, his family and the time and place in which he lived.

William Peter Wheldon 2011

*Jim was a great admirer of the poetry of Housman. He was probably referring to *A Shropshire Lad* part XL which ends:

"The happy highways where I went
And cannot come again"

Notes

Certain abbreviations have been used, notably

LBS – Liverpool Botanical Society

MEC – Moss Exchange Club

BEC – Botanical Exchange Club

LNFC – Liverpool Naturalists' Field Club

RPS – Royal Pharmaceutical Society

Scottish and Welsh place names are spelt as they would have been in Wheldon's time.

William Alfred Wheldon was also known in the family as Bill. Both forms have been used.

Church and Chapel have both been used for Methodist places of worship, although Brook Road is always referred to in the family as church.

ACKNOWLEDGEMENTS

This work would not have been possible without the invaluable help of the following people:

Mark Lawley

Professor Mark Seaward of Bradford University

Botany Staff at Liverpool Museum

Wendy Atkinson and Donna Young of the Liverpool Botanical Society

Sally Whyman, National Museum of Wales, Cardiff

Professor David Clough, Chester University

Mr and Mrs E Coates

David Pimborough

Mr A Finlayson

Mr E Greenwood

Ms D Gay

Raymond Crawford, Walton-on-the-Hill Local History Group

Faye Currie of Pitman Training, Liverpool

INTRODUCTION

In November 1911, as part of its series of articles "Work and Workers" *The Lancashire Naturalist* published a profile of James Alfred Wheldon.

"Of all our workers in Lancashire at the present time, no one is more respected than Mr J. A. Wheldon FLS, who is both a clever and critical botanist and at the same time one of the most affable of men. He holds a unique place in the esteem of Lancashire botanists, and his work has been both original and striking in its value. He is respected by a large circle who have had occasion to consult him with regard to their own work. To these he has always extended a helping hand, and ever with a courteousness and cordiality that have almost persuaded them of his and not their own indebtedness for the opportunity."

Later described as the best all-round botanist of his time in the North of England, (*Flora of the Isle of Man* by D.E. Allen, 1984) James Alfred Wheldon was a competent and well-respected amateur botanist, his reputation extending, according to several obituaries, beyond Britain to Northern Europe and North America, especially for his work on the cryptogams, or lower plants, namely mosses, liverworts and lichens. Although an amateur botanist, his approach to the subject was, like many of his contemporaries, rigorous and professional. His work was recognised by scientific institutions, including the University of Liverpool, and on his death in 1924, his obituaries further attest to the regard he was held in.

He combined his botany with a long career in the prison service for which he received the Imperial Service Medal upon retirement.

He was part of the large growth in interest in botany, and science in general, that was a feature of Victorian and Edwardian times. During the Victorian era the study of Natural Philosophy fragmented into specific disciplines, such as botany, with a more specific and scientific approach being taken and each had many keen and enthusiastic followers. J. A. Wheldon also took an early interest in the new disciplines of ecology and physiological plant geography, which studies the environmental influences on plants, and in phytogeography or geobotany which is concerned with the geographical distribution of plant species.

James Alfred Wheldon was a Yorkshireman, who was compelled by unfortunate circumstances to move across the Pennines to Liverpool where he became a key member of the local botanical community; he will always be best remembered as a Liverpool botanist – hence the title of this biography, "An Eminent Liverpool Botanist" – taken from an obituary in the *Liverpool Daily Courier* published three days after his death.

CHAPTER I

THE WHELDONS OF NORTHALLERTON

James Alfred Wheldon came from a long-established North Yorkshire family of drapers, living for many generations in Northallerton. John Wheldon was born there in 1690; he had two sons, Thomas (b. 1739) and John (b. 1742), the former being the father of twelve children, one of whom, Robert (b. *c.* 1775), was the great grandfather of James Alfred Wheldon. Robert married Mary Morritt, his second wife, in Thirsk in 1801, and in 1803, their son James was born, also in Thirsk. James, later known as James Wheldon Senior, or the elder, married Elizabeth Langtoft; they had ten children, not all of whom survived infancy, as was common in those times. Their fourth child James died soon after his birth in 1834, but the following year another son was born, also named James. He was later to be known as James Wheldon Junior, and was to be the father of James Alfred Wheldon. All the Wheldon children were born in Northallerton, where the family drapery business was established, and after the birth of James there were five more children, one of whom, Robert, died shortly after his birth. Once again the name was re-used for their youngest child, born in 1845. The Wheldon children were Mary (b. 1827), John (b. 1828), Thomas (b. 1830), James (b. 1834), James (b. 1835), William (b. 1837), Elizabeth (b. 1838), Robert (b. 1839), George (b. 1844) and Robert (b. 1845).

Of the older children, little is known. William died in 1874, and by 1881 three branches of the family were living in Station Road, Northallerton where the family business was then based. In the same year James Wheldon Senior had retired and was living in North Terrace, Northallerton with his second wife Ellen, his first wife Elizabeth having died in 1875. He died in July 1889, leaving an estate valued at £985 3s 8d (possibly worth *c.* £75,000 in today's currency).

James Wheldon Junior married Mary Jaques on 23 December 1858; they had eight children. Their first, Annie Elizabeth, born in 1859, died at the early age of fourteen. Their second child, born in Northallerton on 26 May 1862, was named James Alfred. Five more sons and a daughter followed over the next eighteen years. The Wheldon family showed little imagination in the choice of names, with the reappearance of James, Robert, George and William: James Alfred Wheldon's father and grandfather were James, he had uncles and brothers named George and Robert, his second son was William, as was one of his uncles and Elizabeth and Mary appear frequently.

After the birth of James Alfred in 1862, James Junior and Mary had six children, namely John William (b. 1865), George Herbert (b. 1867), Robert Ernest (b. 1870), Lucie Annie (b. 1876), Walter Octavius (b. 1877) and Frederick Thomas (b. 1880). All the children were born in Northallerton and all were christened in the Wesleyan Chapel. In 1881 the six younger children were still living at 116 Station Road, but the oldest, James Alfred, had left home and was by this time starting out on his career as an assistant to a pharmacist in Scarborough. John William was still alive in 1882, but after that date no trace is found.

The Wheldon family trade was wool and linen drapery. By 1861 the business was established in Main Street, Northallerton, where James Wheldon Senior ran it with the help of wife Elizabeth and sons William, George and Robert. At this time James Junior was also in the business, but he had set up home with his wife Mary and daughter Annie in West Row, Northallerton. In 1871, James Senior was retired and the business was being carried on by James Junior, Robert and George from three different addresses in South Parade. James Senior was also in the same road, living with Elizabeth and son William who was then a commercial traveller, possibly for the family firm. In 1881, the family business was described as J. Wheldon and Bros or Wheldon Bros, with James Junior listed as the senior partner. At this time all three brothers had addresses in Station Road. Robert died in 1887, George died of consumption in 1889, and James moved to Bedale in 1885. By 1890, the business was listed as Wheldon & Co. and was run by George's widow Susannah, trading from 16 High Street Northallerton. Susannah, with her unmarried daughter Sarah, were still trading in 1911; since the business was still trading as Wheldon & Co. in 1905, it is assumed that it retained this name hereafter. Susannah, now with the help of her married daughter Emily Oxendale, was still trading in 1930; she died in 1933, aged 87.

By 1891, James Wheldon Junior was living at Market Place, Bedale, with his wife Mary and four children, Lucie, Robert, Walter and Frederick, where he was trading as a draper and milliner. It is not known why he moved to Bedale, but possibly there had been a disagreement with Robert or George, or it may simply have been to develop a new business. At this time, their niece, Martha Jaques, was also living with them, together with a servant. Mary Wheldon died on 12 May 1893 at the Royal Infirmary at Brownlow Street, Liverpool. It is not known why she was in Liverpool at the time of her death; perhaps it was to be near to her eldest son who had moved there in 1891, or perhaps because the Liverpool Hospital had greater expertise (she was suffering from breast cancer). Her body was brought back from Liverpool to be buried at St Gregory's Church in Bedale. A further interesting point concerning Mary's death was that probate was granted to her son Robert Ernest, rather than to her husband. She left just over £1,145, not an inconsiderable sum at that time. James Wheldon Junior lived on for five years after the death of his wife, retiring from business in the spring of 1898. He too died in Liverpool from the effects of influenza, at the home of his son James Alfred at 60 Hornby Road, Walton, on 29 September 1898. He was buried with his wife in Bedale. James Wheldon Junior like his brother George was a freemason and, in common with other members of the family, was a member of the Wesleyan Church. He was prominent in the Temperance Movement; he was also an accomplished musician, much in demand for his playing of the violin, and was a keen member of the Bedale Bowling Club. After his death the drapery business in Bedale was carried on by his son Robert Ernest until his death, when it was run by his widow Elizabeth who was still trading in 1911.

James Alfred Wheldon's other siblings worked in a variety of retail trades. George Herbert was a grocer's assistant in 1891, and by 1901 was a grocer in Newton-le-Willows, North Yorkshire, with his wife Annie and four children. He was still in the same trade in 1911, and died in Leeds in 1933. In 1901 his brother, Walter Octavius, was living with him in Newton-le-Willows; he went on to become a draper in York. Lucie worked in Bedale as a stationer's assistant until her marriage in 1904 to John Jackson who had been a draper's assistant in Bedale, possibly with the Wheldons. They went on to run a fish and chip shop. Frederick Thomas, known as Toddy, was working as a jeweller's assistant in Northallerton in 1901; by 1911, he was in business on his own account as a watchmaker and jeweller in Market Place, Bedale.

Of the two youngest Wheldon brothers, Walter died in York in 1966 and Frederick died in 1960 in Northallerton. Both men outlived their nephews, James Alfred's sons, by many years. To what extent Wheldon kept in touch with his sister and brothers is not now known. They all remained in Yorkshire for the rest of their lives and all survived him and although he made some botanical trips to the county it is not known whether he combined these with a family visit.

CHAPTER 2

PROFESSIONAL AND BOTANICAL LIFE, A FIRE AND A MOVE TO LIVERPOOL

James Alfred Wheldon inherited his love of the natural world from his father, James Wheldon Junior. The latter was mainly a collector of birds' eggs but was also interested in natural history generally and the two would spend a great deal of time roaming the North Yorkshire countryside together and writing up their findings in a journal. There seems to have been a particularly strong bond between them as none of the other children appear to have had such a keen interest in the natural world, although George and Walter may have had some interest when they were young as botanical samples credited to them survive in the Wheldon herbarium today. It would seem that a similar bond existed later between James Alfred and his eldest son Harold. One part of their drapery business was to visit outlying farms and villages, selling their goods from a caravan and these journeys gave many opportunities to study the flora and fauna of the North Yorkshire Moors.

Unusually for an eldest son, the young James Alfred did not follow his father into the family business. He was educated in Darlington at Cleveland College and embarked on a career in pharmacy. This choice of career may have been influenced by his growing interest in botany and the medicinal use of plants. As part of the training for this profession, he spent some time in the early 1880s in Scarborough as an assistant to, and living with James Williamson, a chemist, at 9 South Street, Scarborough. Even during this time he continued and extended his interest in nature by starting a study and collection of seaweeds, the collection surviving until his death over forty years later. During his time in Scarborough he also started to take an interest in fungi, starting a collection of leaf fungi and contributing to *Yorkshire Fungus Flora* and to Grove's *British Rust Fungi.*

After his time at Scarborough, Wheldon studied at the Westminster College of Pharmacy in London and qualified in 1884. At this time the Royal Pharmaceutical Society (RPS) offered two examinations, the minor and the major. The major examination was for business-owning pharmacists, and the minor was mainly for assistants. James Alfred Wheldon passed the minor examination in December of that year, becoming a registered chemist and druggist. Before sitting the examination, candidates would undertake a period of apprenticeship (as Wheldon did in Scarborough) and study by correspondence course or at a college of pharmacy. Until 1898, chemists and druggists were ineligible for membership of the RPS, but there is no evidence that Wheldon ever applied for membership after that date. At the time of his examination he was living at 9 East India Dock Road in East London, but by 1885 he had returned to Yorkshire to his family and lived in South Parade, Northallerton, and later in the year in Market Place, Bedale. By February 1887, when he married Catherine (also known as Caroline) Simpson, he was living and working in York and from 1886 had been practising as a pharmacist at 20 High Ousegate, York. He lived on the premises and settled down to married and family life. His two sons, Harold and William, were born in York in 1888 and 1889 respectively. Given his qualifications, it would seem unlikely that he owned the business although chemists and druggists who then became proprietors could become 'associates in business' of the RPS. However, according to the profile of Wheldon in *The Lancashire Naturalist* the business "opened out very successfully".

His new business and family did not curtail Wheldon's interest in the natural world. Botany was his main area of interest but he still maintained an interest in other areas including Oology (birds' eggs) and Entomology (insects). In addition to a large herbarium and an extensive library, he had built up collections of eggs and moths and butterflies and also developed skills as a taxidermist, producing specimens of birds which he displayed in glass cases. According to one obituary he maintained an aviary and in addition, he took an interest in shells, but generally he was beginning to concentrate on botany. Until about 1890 this interest was chiefly in the phanerogams (flowering plants), but after this date he developed an interest in bryology, but certainly not to the exclusion of other types of plant. From about 1886, Wheldon started to write articles on natural history subjects, his first published works appearing in the *Yorkshire Chronicle* as a series named "Wild Flowers of the Week". The *Yorkshire Chronicle* was a weekly publication first published in 1855, becoming a daily from 1888; these articles were soon followed in 1889 by a number of contributions to *Hardwick's Science Gossip*, a "monthly medium of interchange and gossip for students and lovers of nature". Wheldon's contributions illustrate his wide range of interests, from flowering plants to mosses, and to even the colouration of birds' eggs. Incidentally this journal also published a letter in 1895 from the Rev. C. H. Waddell that lead to the founding of the Moss Exchange Club with which Wheldon was closely associated from its foundation until his death.

From the age of about fourteen, Wheldon took an active interest in scientific clubs and societies, and by this date

(1876) was already a member of the Botanical Exchange Club. His first recorded contribution to the Club was from Romanby, close to his home in Northallerton, in that year.

On 31 December 1890, the Wheldon family suffered a major setback when a fire broke out at 20 High Ousegate. Damage was extensive and in addition to his stock in trade much of Wheldon's library and collections were lost. The cause of the fire is not known and no contemporary account of it has been found, but fortunately no one was injured (although the birds in his aviary perished) and it is evidence of Wheldon's resilient nature that he immediately set about rebuilding his collections.

The fire must have been a traumatic experience for Wheldon and his wife as their two young sons were only one and two years of age at the time. The story within the family is that there was no insurance in place, but given that Wheldon was an obviously intelligent man this seems unlikely. However, having lost both livelihood and home the family were forced to move. They lived for a short time in 1891 at Langham Street, Ashton-under-Lyne, and he gained employment as a commercial traveller. During the short stay in Ashton daughter Doris was born prematurely and was not expected to survive. Despite the upheaval and circumstances of his daughter's birth, Wheldon used the opportunity to make some botanical observations in the area. At some time before 1893, he worked on liverworts with Benjamin Carrington (1827-1893), the medical officer for health in Eccles. Later in 1891, Wheldon obtained the post of pharmacist at Walton Prison in Liverpool, a post he was to hold for the next thirty years. The family moved initially to 9 Chelsea Road, off Walton Vale Road and then to 60 Hornby Road, Walton, one of the houses adjacent to the prison and built for prison staff, and where all of Wheldon's neighbours were employed in various capacities at the prison. This was to remain his home until his retirement in 1921. Upon his retirement, he was awarded the Imperial Service Medal, a decoration given to those of low or middle rank with long service in public life. (An Imperial Service Order was given to those of higher rank.) It is probable that Wheldon's intense interest in botany restricted his professional career as he seems never to have sought to upgrade his professional qualifications or to look for higher status.

The Wheldon family's move to Liverpool in 1891 must have been something of a culture shock for them. James Alfred and Catherine had been brought up in Northallerton and Bedale respectively, both small Yorkshire towns essentially rural in nature. Northallerton had a population at this time of about 4,000, and Bedale was even smaller. Both towns would have had many long established families, such as the Wheldons and the Simpsons, and people would tend to know each other. In contrast, Liverpool had a population of around 700,000 with inhabitants drawn from many ethnic and cultural backgrounds. Although he had lived in Scarborough and York (both comparatively small towns) and briefly in London, the move to Liverpool must have been something quite new for Wheldon. Liverpool at that time was at its zenith, claiming proudly to be the second city of the Empire, being an exciting and vibrant place full of Victorian and Edwardian confidence. It was Britain's largest exporting port and departure point for the great transatlantic liners of the Cunard and White Star lines. It was from Liverpool that the Lusitania and Mauritania made their maiden voyages in 1907 and it was the home port of many other great ships. In addition it was an industrial city, with tanneries, sugar refineries and various mills and foundries as well as the substantial Hartley's jam works only a few hundred yards from the Wheldons' new home in Chelsea Road. The growth of the city continued during Wheldon's time with the expansion of the north docks not completed until after his death.

Tram journeys into the city centre would have taken him through Kirkdale and along Scotland Road, past row upon row of terraced housing, famously with a pub on every corner, and infamous for their poverty and appalling conditions. But, like all Victorian cities, there was a great deal of wealth to contrast with the poverty.

There was a thriving social and cultural life in the city with many clubs and scientific societies which were not only serious scientific institutions but staged social events as well, giving opportunities to people such as the Wheldons to both pursue their interests and develop friendships. He was a member of the Liverpool Naturalists' Field Club; he also served on the committee of the Associated Learned Societies of Liverpool and was active in the formation of the Liverpool Botanical Society. In addition, he was associated with Liverpool University, where he worked in its herbarium, for which he was awarded an honorary MSc. in 1922. The city also provided other opportunities for his studies, including the Botanic Gardens in Wavertree (established in 1803). The Liverpool Naturalists' Field Club dated back to the 1860s and its membership had a wide range of interests including botany. Wheldon was a member from 1901 until 1906 and for three years from 1903 served on the committee, a familiar role as previously he had been a member (and secretary) of York & District Naturalists' Field Club. During this time the Liverpool club had around 200 members and their field meetings usually attracted around twenty-five participants. As the club included botany among its activities it is not clear why a separate Botanical Society was formed in 1906; perhaps the botanists wanted their own organisation, or perhaps there had been some form of disagreement. Wheldon ceased to be a member of the LNFC when the LBS was formed and certainly some of the founding members of the LBS were former members of the LNFC. Why he did not retain his membership is not now clear. The year 1905 saw the lowest-ever attendances at LNFC meetings, but that was put down to poor weather rather than any reason that might have led to a split.

During Wheldon's time in Liverpool some of the iconic structures in the city appeared, which emphasised the wealth and importance of the city. The pioneering overhead railway was completed in 1893, passing on its way the 'three graces' which provided Liverpool with a world-famous and instantly recognisable waterfront, namely the Dock Board Building (started in 1907), the Royal Liver Building (opened in 1911) and the Cunard Building (built during the First World War). On his train journeys into Liverpool, Wheldon would have seen the construction of the tobacco warehouse at Stanley Dock in 1900 – at the time the world's largest building, constructed with 27 million bricks – and during his life would have seen, on the other side of the Mersey, the construction and demolition of New Brighton Tower, around 50 ft taller than Blackpool Tower, started in 1896 and taken down in 1921. The Victoria Building of University College Liverpool opened in 1892; the foundation stone for the magnificent Anglican cathedral was laid in 1904; the Adelphi Hotel was rebuilt in 1912; and in 1892 Everton Football Club moved out of Anfield to Goodison Park, one of the country's first major football stadia. The famous Philharmonic Dining Rooms and the Vines public house (the Big House) were built and rebuilt in 1898 and 1907 respectively. This then was the Wheldon family's new home city and they soon settled in and became respected members of the community. Wheldon's name as a botanist became associated with Lancashire generally and Liverpool in particular.

Liverpool and Lancashire with their attendant industry, all coal-fired, might on the face of it seem unproductive ground for botanists. Writing just after the turn of the 20th century in his book *The Flora of West Lancashire*, co-authored with Albert Wilson and published in 1907, Wheldon noted the effect, especially upon mosses and lichens, of the pollution caused by 'coal combustion'. Even diluted by distance, he said, it was carried on prevailing winds and had an effect on the health and distribution of these plants. This observation was not a particularly new one at the time, but it was especially relevant in the local area. He returned to this theme in some detail a paper he wrote with W. G. Travis in 1915 entitled *The Lichens of South Lancashire* in which the authors noted the particularly high levels of pollution in industrial South Lancashire (lichens are especially sensitive to pollution). To illustrate the point, Valentia in Southern Ireland was taken as a base point for air pollution by sulphates with a value of 100, with inland English counties having an average value of 202.2. Liverpool had a value of 1,450.2 and nearby St Helens a value of 1,215.8. Concern about pollution is not necessarily a recent thing, although perhaps its link with global warming was not as widely recognised in 1915. In the paper the authors looked forward to the day when pollution would be reduced and interestingly saw the internal combustion engine as a step in the right direction. They could not have known that this would also add to the problem. Industry was not wholly to blame, however, and the authors recognised domestic chimneys as making a significant contribution. The problem of pollution in the Mersey region remained until recent times, with smogs still occurring in the 1960s and the development of the petrochemical industry adding to the problem.

Similarly, increasing urbanisation was also giving concern to botanists and a reviewer of *The Flora of West Lancashire*, in the journal *Nature*, was fearful for the botany of such areas due to expanding 'industrial enterprises'. Despite this, however, botany was thriving in the area, and in the same book the authors noted that "Lancashire has always been remarkable for the number of its botanists"; thus Wheldon had clearly moved to an area with every opportunity to develop his interest. This of course was long before the advent of Merseyside as a county and Liverpool was firmly within Lancashire. Incidentally, it has been recorded that industrialisation provided some bonuses for the botanist; for example, the warm water discharged from industrial premises into local canals and watercourses provided a suitable environment for some unexpected species.

This subject of industrial activity and the effect on botany was clearly one that interested Wheldon. In 1909 he wrote an article in *The Lancashire Naturalist* titled "On the Influence of Railways on the Local Flora" in which he summarised the positive and negative effects, and interestingly suggested the establishment of nature reserves where plants could survive away from the effects of industry.

"In dealing with the Flora of a densely-populated district, special attention must be given to the effects produced by the alteration of natural conditions through man's handiwork. Canals have a marked influence on the preservation and introduction of aquatic plants. Quarries, walls and even the roofs of buildings play their part in affording habitats for plants that would probably not occur in the district under strictly natural conditions. Second only in importance to the operations of agriculture and horticulture should be ranked the railways. Naturalists often have to deplore the indirect results of the network of iron roads which has been spread over the length and breadth of this country over the past one hundred years. Only by this expansion has the phenomenal growth of cities and industries been possible, and their effect on botanical features of the country has been most deleterious. Few counties show more conspicuous example of this than South Lancashire. The Flora has suffered severely, not only from the rapid covering of many choice plant habitats by buildings and roads, but also by an increase in atmospheric pollution by smoke. In addition, places once inaccessible and secluded have been thrown open to troops of excursionists, with disastrous results to Flora and Fauna. Especially have all these factors exercised their evil influence on the Flora of littoral Lancashire, the seaboard having become through enhanced railway facilities a mere suburban annexe to the large inland manufacturing centres. As the growth of commerce has been chiefly instrumental in destroying an almost unique tract of country, a very paradise for naturalist, the latter should look to those who have grown rich for aid in the preserving of some of the remnants which are left. Such attempts have met with success on a larger scale in America and Africa. It should not be impossible to obtain from the Napoleons of trade a sufficient amount to purchase a few acres of sand dunes, and maintain them in their natural condition, to delight with their treasures future generations on Nature lovers.

Every cloud is said to have a silver lining, and even our railways, from a botanical standpoint, are not entirely malevolent in their effects. To some extent they have been conservative instead of destructive. The embankments, cuttings and ballast heaps afford refuge to many interesting plants. Orchids, cowslips, primroses, burnet and ox-eye daisies clothe the banks in many places where they have been long banished from surrounding fields. On the outskirts of large towns, secure from grazing cattle, the plough and the jerry builder, many plants once common now lurk by the railway only, and it is useless to look for them elsewhere. In addition a variety of introduced plants thrive and maintain their hold from year to year. In the vicinity of Liverpool such immigrants are seen in *Geranium pyrenaicum, Trifolium incarnatum, Senecio viscosus, Coronilla varia, Sisymbrium pannonicum, Enothera biennis*, and many of less permanence. The little toadflax, which I used to gather as a cornfield weed in Yorkshire thirty years ago, has become a confirmed railway traveller in Lancashire, and follows the railway through the county.

Perhaps the most interesting refuges for plants are certain small plots of land purchased with a view to further extensions, and then allowed to lie idle until wanted, and especially small triangular areas enclose by the railways and some of our smaller junctions. Here one may still find *Orchis incarnate, Salix repens* and other willows, brambles, roses and even the little adder's tongue fern. A few such situations contain pools of interesting aquatic plants including Chara, surrounded by thickets of Iris, Reed Mace, Sparganium, Grasses, Sedges and Horsetails.

Not only are the higher plants the only ones which take advantages of these refuges provided by the companies. Fungi are plentiful, not only the larger kinds, but many interesting micro species also. On a bank near my house the Ladies Mantle is orange with *Uromyces intrusa* and the heads of the Yellow Goatsbeard are purple with *Ustilago receptaculorum.* This latter does not appear to affect the Purple Goatsbeard, a rare species which has for more than half a century found refuge on the banks near Liverpool. In the boggy situations mentioned above, many mosses thrive in great luxuriance such as Bryum *bimum, Fissidens adiantoides, Hypnum stellatum, H. Polygamum, H. Aduncum, H.Sendtneri, H.revolvens, H.scorpioides,* and even the rare little *Swartzia inclinata.* In drier places we may find *Weissia microstoma, Fissidens viridulas, Polytrichum gracile* and many commoner species. It is hoped botanists in other parts of the country will not overlook such habitats when exploring their own districts. A Railway Flora of South Lancashire, if properly worked out, would prove to be interesting reading, and would supply a good many surprises and some matter for speculation as to the origin of the component parts."

Wheldon's new home in Walton was then on the fringe of the built-up and industrial areas and not as developed as it is today, but he lived to see the area between there and Bootle start to develop; he also witnessed the building of the Orrell Park estate to the north of Hornby Road, and indeed moved there himself. Living in Walton did give fairly easy access to the agricultural areas of south west Lancashire and to the dunes on the coast, both regular destinations for botanical trips. His change of employment would also have been something of a shock, from a chemist's shop in relatively genteel York, to the harsh environment of a prison with a thousand inmates, many of whom would have been serious and dangerous criminals with whom he would have had regular contact.

In 1896, James Alfred Wheldon was one of twenty-three founder members of the Moss Exchange Club (MEC), and he played a prominent part in the affairs of the club for most of the rest of his life. Wheldon was a member of three Exchange clubs and at times acted as distributor in them all. These clubs were a Victorian and Edwardian phenomenon, providing both a platform for the study of botany and also a social network. They enabled members to build up their own collections, or herbaria, and also gave access to experts (referees) to help them identify specimens and increase their knowledge of the subject. An accusation often levelled at Victorian and Edwardian botanists is that they tended to over-collect and exchange clubs may have been partially responsible for this as members would collect samples of plants not only for their own use, but also to distribute to others for their own collections.

Wheldon held three main positions within the Moss Exchange Club. In 1900 and 1901 he was the distributor, receiving bulk samples from members for re-distribution to other members of the club. For most of the period from 1903 to 1919 Wheldon was a referee for the club, giving advice and helping to identify samples. In later years the club moved to a system of specialist referees, so from about 1914, Wheldon was referee for two rather difficult groups of mosses, namely the *Sphagna* and *Harpidia*.

From 1904, Wheldon was the honorary treasurer of the club and appears to have held this post until at least 1919. The club's 1914 report refers to him as our "skilful treasurer", but over the years it would seem that the MEC was not without financial difficulties. Wheldon wrote in 1919 to Arthur Dallman, a friend and fellow member, about the difficulty in meeting printing costs (which he considered too high) and suggested that subscriptions would have to rise, and in 1911 when the club was preparing its *Census Hepatic Catalogue*, Wheldon offered to pay for its production as funds were limited, hoping to reclaim his money from subsequent sales. In most years, Wheldon contributed specimens of both mosses and hepatics for distribution, although the wartime years 1913-1916, as would be expected, were his least productive in this respect.

James Alfred Wheldon was a very thorough and precise botanist. He was very keen that the specimens sent into the club were correctly presented and labelled. In the early years of the club, standards were perhaps not as high as he would have liked. He wrote in the MEC report of 1901 that "specimens are improved" but went on to make some further observation, namely that "specimens are not carefully dried" and that there was a "paucity of material in some packets". He also reflected that insufficient specimens were sent to the referees. Things must have improved as by 1913, when he observed that specimens were excellently prepared and accurately labelled. The First World War seems to have had an effect on the collections and samples and Wheldon noted in the 1916 report that, for example, it was difficult to get paper of the right quality upon which to mount samples. (The paper used is one of the determining factors in the life of samples in herbaria.) He went on to give advice in 1917 on the preparation of *Sphagna* samples: they should be "spread thinly and dried with gentle pressure". Wheldon noted that generally continental samples were better prepared.

In 1916, Wheldon had agreed to prepare a new *Sphagnum* Catalogue for the club. There were at that time two systems of naming *Sphagna* and it was proposed that a new system based on Warnstorf's *Sphagnologia Universalis* be used. Wheldon was also a member of both the Lichen Exchange Club and the Botanical Exchange Club. The former was a relatively short-lived organisation, but the Botanical Exchange Club became the Botanical Society of the British Isles, and the Moss Exchange Club later became the British Bryological Society, both of which continue to flourish.

Wheldon did not hold any major office in the BEC, but he was a prolific collector of specimens (Appendix 3) and a contributor to discussions; for example, in 1912, together with two fellow members H. J. Riddlesdell and E. S. Marshall (both clergymen), he prepared a report on the identification and taxonomy of the rare Radyr Hawkweed.

Wheldon was a founder member of the Liverpool Botanical Society (LBS) in 1906 and continued to be an enthusiastic member for the rest of his life, regularly attending indoor and field meetings, some of which he led, and on many occasions lecturing to the Society. The foundation of the Society can be traced back to discussions held at 60 Hornby Road in 1905 or 1906, which led to Arthur Dallman, the first secretary, circulating a letter to interested people. The first preliminary meeting took place on 26 April 1906 at the Common Hall in Hackin's Hey, across the road from Liverpool Exchange Station. On 21 May, the first full meeting took place at the same venue and officials were chosen. Wheldon became the first vice-president, jointly with the Rev. William Wright Mason, A. A. Dallman was secretary and the Rev. Samuel Gasking was the first president, after Wright Mason had turned the office down. The first treasurer was F. J. Routledge, a headmaster living in Anfield Road.

Over the years, Wheldon held various positions in the Society. He was president in 1909 and 1910 and a vice-president on several occasions. The Wheldon family were great supporters of the LBS; apart from Wheldon himself, Catherine was a founder member and Harold was a prominent member until his marriage and move to Warwickshire in 1911, after which he still kept up his membership until the time of the First World War. To complete the family connection, Doris was a member of the LBS until at least 1922, even though for some of that time she was in the Far East. Only William appears to have had no interest in the Society, or, it would seem, in botany. Wheldon and Harold made significant contributions to the Society, as did Catherine who took a leading role in its organisation and for several years was a member of the committee for the annual soirée, an evening with some botanical discussion, but mainly a social affair, which she hosted in her husband's presidential years.

Catherine not only supported her husband in his roles in the society, but also made some contributions of her own. In 1910 she moved an amendment to rule 10 of the Society, which declared that all indoor meetings would be on Mondays. She quite reasonably suggested that the day should vary and the motion was carried. In 1911 Harold was to have been the Society's librarian but on his move to Ryton on Dunsmore, Warwickshire, he had to relinquish the post. His mother stepped in and became the temporary librarian. Her sudden death in September 1915 shocked the Society and at the following meeting on 22 October (a Friday!), the president made sympathetic reference to "the loss sustained by the Society on the death of Mrs Wheldon, a founder member of the Society who had served on the Council for many years and had assisted in developing the activities of the organisation". The huge personal loss did not prevent her husband exhibiting botanical samples at the next meeting a month later.

During Wheldon's lifetime, the LBS usually had around 100 members, and most meetings attracted about a quarter of the membership. Members were by no means all from Liverpool, with some coming from other parts of Lancashire and Cheshire, which is perhaps some indication of the status of the Society. The following extracts are taken from local newspapers and give an idea of the activities of the Society and of Wheldon's involvement.

> Liverpool Botanical Society – The fourth annual soiree and convesazione was held at the Palatine Cafe on Saturday night. After a reception by the president and Mrs Wheldon, and a short address by Mr Wheldon, a concert was contributed to by various members and friends. The items of this programme were excellently rendered by Miss E M Coates, Mrs Davies Miss Annie Hall, Miss Kennison, Miss Ada Whitehead, Mr J A Kennison (accompianist) and Messrs Davenport and Evans. For the nonce, the scientific aspect of the society'soperations was unobtrusive. An interesting feature due to the collaboration of various members was a large and unique exhibit of February flowers. Fresh flowering examples of outdoor plants only, British and acclimatised were shown, greenhouse productions being excluded. Altogether eighty-two different species were on view and in flower, a very creditable total having regard to the time of year. These were contributed by Miss E Bray, Miss Agnes Fry, Miss E G G Hill, Miss L N Peacock and Messers

R H Day, Hillman, J A Wheldon and A A Dallman. Dr Ellis showed some extremely realistic coloured drawings of fungi which aroused general admiration and an interesting collection of plants of economic importance, contributed by Mr W Hackett of the Botanic Gardens, also called for note.

Liverpool Echo, 21 February 1910

Wheldon's contributions were probably grown on his allotment in Orrell Park

The ordinary monthly meeting of the Liverpool Botanical society was held at the Botanical Lecture Threatre of the University. Mr J A Wheldon FLS (president) was in the chair. A number of interesting exhibits and communications were received. Miss E G G Hill sent some living specimens of the peculiar little ranunculaceous plant, myosurus minimus, from Portbury, Somerset. This species was supposed to have long become extinct in Somerset until lately rediscovered growing in abundance in one station by Miss Hill in company with Miss L N Peacock of Bristol. Miss E Bray contributed a number of specimens to illustrate the variations of ranunculus tripartitus. Dr J W Ellis gave some account of some fungi observed at the society's recent field meeting at Rossett, and exhibited a number of coloured drawing of local fungi. The president showed some fruiting examples of the splachnun sphoericum, a curious subalpine moss which usually occurs in decaying animal matter on moorland localities. Mr W G Travis announced that a moss which had been obtained on the South Lancashire sand dunes, and submitted to Mr H N Dixon, had proved to be ceratodin conicus. This plant, which is perhaps a form of the polymorphic ceratodin purpureus, is, according to Mr Dixon, identical with the plant which occurs under very similar conditions on sand dunes at Dunkirk. Various exhibits and communications were also contributed by Miss L N Peacock, Mr R W Sibbald, Mr Harold J Wheldon, Mr R H day and other members. The feature of the evening was a paper by Miss F M Thomas entitled 'Notes on Radnorshire Plants'. In this valuable contribution the authoress succeeded in giving a useful account of the vegetation of an area which has hitherto received all too little attention from botanists. On the proposition of Mr R H Day, seconded by Mr Harold J Wheldon, a cordial vote of thanks was passed to Miss Thomas.

Unidentified local paper, possibly the Liverpool Echo, 11 May 1910

The members of the Liverpool Botanical Society, accompanied by members of the Dyserth Field Club, visited Formby Sandhills, where under the leadership of Mr J A Wheldon FLS the dune flora was carefully examined. Those who remembered the arid stretch of sand visible here a quarter of a century ago brought to notice the remarkable change which the terrain has undergone, the dwarf silky willow now covering acres of ground, while the pine plantations afford a glimpse of foreign scenery which has earned one corner the name of Little Switzerland. Several capital examples of the sea blackthorn displayed massed of orange coloured berries and the ever attractive grass of Parnassus carpeted the turf with its delicately veined white flowers. The winter-green still persevered in flower and at one point a perfect garden of unusually fine toadflax delighted the eyes of the party. A feature of the excursion was the abundance of fungi noted, many of the presenting a singularly effective colour scheme. Mr A A Dallman FCS named a rare puffball, which occurred in the damper localities, and on the willows and poplars quite a large variety of galls was observed, the most interesting being the spiral pemphtiole of the leaf and gives rise to the spiral form presented a most fascinating object for research, and it was gratifying to observe that this department of knowledge, affecting so closely the operations of farmers and gardeners, is receiving more of the attention it deserves. The duties of recorder were efficiently performed by Miss Olive Bangham.

Liverpool Courier, 25 September 1917

(The above-mentioned pine forests are probably those that today form the Red Squirrel Sanctuary at Formby)

The tenth annual soiree of the above society was held on Saturday evening, when an unusually large number of members and friends assembled to hear an excellent programme of music rendered by Miss Rose Isaacs LRAM, Miss Grace Hughes, Miss Olive Bangham, Mr W Smith, and Mr E Horton. An item, which was much applauded, was the poetical account of the society's proceedings, compaosed and recited by Miss E Warhurst, LLA.

Some exceedingly interesting botanical exhibits were on view, including rare exotics and fruiting examples of native species, among the latter being a particularly fine collection sent by Miss E Bray of Hailsham. Miss S J Shoobridge furnished a series of exquisitely finished silver ornaments made by handworkers in India and Mr J A Wheldon FLS displayed a most instructive and interesting collection of sphagna. These bog mosses, on which Mr Wheldon is an acknowledged authority, were named according to the latest pronouncements of Dr Warnsdorf, the microscopic details being carefully drawn for each species. Although the lighter side of science was prominent there was ample opportunity for serious students to add to their knowledge, and the facilities were fully availed of.

Mr A A Dallman FCS and Mrs Dallman, in the enforced absence of the president Mr W G Travis, performed the duties of host and hostess with much acceptance.

Liverpool Courier, 23 October 1917

A meeting of the Liverpool Botanical Society was held on Monday evening, Mr A A Dallman presiding. Mr J A Wheldon exhibited specimens of sphagna from Essex; Mr W G Travis showed forms of the common pear tree, of which several were from stations in which it is regarded as native. The president submitted a choice selection of sprays of flowering species from Grosvenor Park, Chester including jasmines, hellebores, spurges, and fruiting specimens of aucuba, this condition being formerly unknown in England. Mr Euston Payne showed an interesting group of mosses from the Doncaster district.

A paper was read by Dr W A Lee on 'The scope and functions of a botanical society'. He said they should not only study those well-known organisations such as the Linnean Society, but they could derive much benefit from a close examination of the work done by the old Lancashire botanical societies. These were prior to the Linnean Society, having been in some instances formed before the last quarter of the 18th century.

Liverpool Courier, 27 February 1918

(Presumably the *Sphagna* from Essex came via the Moss Exchange Club. It is unlikely that Wheldon visited that county. The final paragraph is interesting in the light of Wheldon's comments in his *Flora* about the number of botanists in Lancashire. The county seems to have had a long tradition of an interest in botany.)

The indoor and outdoor activities of the LBS were regularly reported in the local press and Wheldon is regularly identified as a speaker or leader of trips, some of which were held jointly with other societies, such as the Chester Society. The reports also mention various members of the society, perhaps one of the most interesting being Dr J. E. W. McFaul, later Professor of Forensic Medicine at Liverpool University and a prosecution witness at the famous (or infamous) Liverpool murder trial of William Herbert Wallace in 1931.

In 1907 or 1908, a committee was established by the Society to prepare a *Flora of South Lancashire*. Wheldon sat on the committee, as did Harold until his move south in 1911, the latter with special responsibility for fungi. Harold also sat on the Society's Council. It is recorded in the Society's *Proceedings* that the progress of this *Flora* was slowed down by the First World War, particularly in terms of railway restrictions which prevented some fieldwork.

Every year the Society held regular meetings. These were generally indoors in the winter months, but during the summer outside or field meetings were held. These tended to be fairly local but often ventured a bit further afield, for example, to North Wales. As an example, the following field meetings took place in 1911 at these destinations: Llanfynydd; Frodsham and Helsby; Hatchmere and Delamere; Rufford; Freshfield; and Llanfihangel Glyn Myrfyr. Today these trips would present few problems, being a relatively easy car or coach journey, but in 1911 they would be rather more difficult, and some are a real tribute to the enterprise and enthusiasm of members of that time. Llanfihangel Glyn Myrfyr is a fairly isolated village in North Wales. The only way to reach it would be by train. On a good day with fast trains Corwen could be reached in about 2½ hours; a further twenty-minute journey would take them to Derwen, the nearest station to Llanfihangel; this line, however, had a very sparse service with only about six trains a day. Then there would be a 5-mile walk to the destination. On the day of the LBS trip, the train to Derwen was missed and a wagonette had to be hired to complete the journey! Another interesting field trip was made in September 1913 to the Conway Valley, starting with a train journey to Llanrwst. The day would have involved a round trip of about fifteen miles, involving a climb of about 600ft. It also illustrates that the trips were not purely botanical, since opportunities were taken to look at places of historical or cultural interest.

LIVERPOOL BOTANICAL SOCIETY
VISIT TO GEIRIONYDD LAKE

The eighth field meeting of the session was held on Saturday, and took the form of a whole-day excursion to a district in North Wales hitherto unvisited by the society. On arrival at Llanrwst the party crossed the Conway, and reaching Caernarvonshire, proceeded along the road by way of Gwydir house, the latter being the ancient seat of the family of Wynne and the centre of many historical associations. Ascending by woodland paths, many charming views of the Conway valley were obtained on route. At one point a glimpse of the cataract Rhaiadr Y Parc Mawr was obtained. The very local *Thlaspi alpeatre* was found in some quantity, and the bramble *Rubus anglo soxonicus* was noticed in various parts of the woods. Another plant rarely encountered on the society's excursions was the water purslane (*Peplis portula*) which was quite plentiful on the margin of a small artificial lake.

On reaching the small mountain village of Llanrhychwyn a visit was paid to the quaint old church which is probably the oldest and most primitive of its kind in this part of the country. The dainty ivy leafed bellflower was noticed and the type form of the graceful Parnasia palustris or grass of Parnassus was keeping it company. Llyn Geirionydd, noted for its associations with Taliesin the Welsh poet, who lived here in the sixth century, was next reached. Some of the party continued over the rocky extremity of the peak Mynydd Deullyn to Llyn Crafnant. The bog myrtle and three species of insectivorous plant occurred hearabouts, and this lake appeared to offer much greater variety botanically than the former.

After tea at Llyn Geirionydd, the return was made by a different route through charming scenery by way of Trefriw. *Clematis vitalba* and the form of *cirsium arvense*, which has been named setosum, were found on the way to the station. This was undoubtedly the most enjoyable and interesting field trip of the season.

It is not recorded whether Wheldon was present on these trips into Wales but as a very keen member of the society it is likely that he was.

In 1901, Wheldon was elected a Fellow of the Linnean Society of London, a prestigious society "devoted to the cultivation of the science of natural history in all its branches". Membership was by election, an indication of the status Wheldon had achieved. He was elected on 17 January of that year, being nominated by six members who included Albert Wilson and Hugh Neville Dixon, both of whom were life-long friends and colleagues. At the time of his election he was noted as an expert on mosses, but his main areas of interest were given as botany and ornithology, the latter being an interest he held throughout his life. He apparently withdrew from the Society in 1918, but in 1923, after his retirement, he was made an Associate of the Society, an honour bestowed on him for his original botanical work, being accorded only to a limited number of scientists.

All through his life he continued to contribute articles to a range of scientific journals and publications. On two occasions during his lifetime, the British Association held its annual meetings at Southport in 1903 and at Liverpool in 1923; on both occasions Wheldon contributing to the *Handbook* specially produced for the visits. The British Association was founded in 1831 as an alternative to the Royal Society, which was viewed by some in the scientific community as elitist. It held its annual meetings at different locations around the country and was concerned with many aspects of science, including botany. Wheldon's 1903 contribution was on the mosses and liverworts of the Southport district, but in 1923 he was a member of the committee that helped to arrange the Association's visit to Liverpool, and vice-president of the 'K Section' which dealt with botany.

Wheldon's best known publication is *The Flora of West Lancashire*, published in 1907, which he co-authored with Albert Wilson. He first met Wilson in about 1884 and the two became life-long friends, spending a great deal of time in each other's company on botany field trips and on holidays. Indeed, records of botanical trips suggest that Wheldon may have spent more of his leisure time with Wilson than he did with his own family!

Published Floras are descriptions of the botany of given areas, their geographical limits usually based on the work of Hewett (sometimes noted as Hewitt) Cottrell Watson (1804-1881), who instituted a system of vice-counties (sometimes subdivisions of counties) that is still used today. Watson, a Yorkshireman by birth, was a botanist, evolutionary theorist and phrenologist, and his help with the *Origin of Species* was acknowledged by Darwin. His major published works were on topographical botany through which he established the vice-county system of recording. Of the 112 vice-counties delimited for England, Scotland and Wales, West Lancashire (actually mid-Lancashire) was vice-county number 60. At this time many vice-counties had published Floras, but both West Lancashire and South Lancashire, number 59, lacked these. Originally Lancashire had been split into three, the third vice county being Lake Lancashire, but this became incorporated in Westmorland and Furness. When or why Wheldon and Wilson decided on West Lancashire is not known; on the face of it, South Lancashire would have been easier for Wheldon, but possibly it was Wilson's choice as he had family connections in Garstang and had probably started collections in the area at an earlier date. Certainly some of Wilson's collections noted in the book predate any of Wheldon's, whose first record is from 1896, whereas collections recorded by Wilson go back to 1881. It is also possible that West Lancashire would prove more diverse and interesting as it had a wider range of habitats than South Lancashire, especially to the east, that had been comparatively neglected by botanists at that time. The east of the vice county tended to be higher and wetter as compared to the drained, cultivated land to the west, especially in the Fylde area. The importance of this was illustrated in the book in a chapter on the distribution of mosses, hepatics and lichens, an area of special interest to Wheldon and one that would probably define his botanical career. An analysis was made of the distribution of these cryptogams in the three divisions, north, east and west of the vice-county and note was made of the factors influencing this distribution, including temperature and more importantly rainfall. The average annual rainfall at a number of places in the vice county was taken over the period from 1891 to 1900; as an example, results from three locations are as follows:

Arkholme	North division	46.55 inches
St Michael's on Wyre	West division	36.25 inches
Stonyhurst	East division	47.95 inches

The authors also suggest that the fell country in the far east of the vice-county, where they had not had the opportunity to obtain results, would be even wetter with rainfall of around seventy inches per year – this assumption was based on results obtained from just over the Yorkshire border. To prove this point, in 1906 a rain gauge was placed at an altitude of 1,630ft on Fairsnape Fell and this produced a reading of 83.02 inches. It was further pointed out that the humidity in the fell country to the east was also much higher, giving favourable conditions for the mosses and hepatics which, it was noted, were abundant and luxuriant.

Area	North and east	West
Type of country	Chiefly hill country	Chiefly lowland
Sphagna	27	19
Mosses	241	64
Hepatics	91	29
Lichens	262	40
Total (species)	629	152

Table 1. Flora of West Lancashire p. 108

Altitude (ft)	Sphagna	Mosses	Hepatics	Total
0-100	3	29	3	35
100-300	1	35	9	45
300-500	1	33	8	42
500-1000	5	62	26	93
1000-1500	11	56	21	88
1500-2080	10	83	32	125

Table 2. Flora of West Lancashire p. 109

The influence of altitude on distribution was also considered (see Table 2), although the authors were not able to include the lichens in this analysis. This may begin to explain why, after the publication of the *Flora*, Wheldon's botanical trips, given his interests, tended to be to noticeably hillier, and wetter, regions such as Scotland, the Lake District and the Isle of Man.

As mentioned above, for the purposes of the *Flora*, the vice-county was divided into three (north, east and west) and each of these areas was subdivided into a total of eight districts to provide a more precise geographical location for a particular specimen. The book had a conventional layout with chapters on the topography, meteorology and climate, followed by taxonomic lists. The *Flora* was a major work of almost 500 pages based on about ten years' research. It covers an area from Preston to Kirkby Lonsdale and from Blackpool almost to the Yorkshire border. Study of this area, at some distance from the homes of the authors, involved a great deal of travelling which had to be fitted in around their work commitments. The authors refer in the preface to "wearisome railway journeys" and admit that lack of time meant that some library research had not been done; they preferred to spend what time they had on field research. The book was published in 1907 and there appear to be two editions, both in that year. One was published by the publishers and booksellers Henry Young and Sons in Liverpool, and the other was published by Wheldon and Wilson themselves and printed by V. T. Sumfield of Eastbourne; it was available from the authors at their home addresses for 12/6d. Publishing the book themselves must have involved some considerable financial risk to Wheldon and Wilson.

The fairly extensive description of the topography of the area suggests that the authors covered most of it personally, no mean feat given the nature of the terrain, especially in the east. Wilson was particularly interested in meteorology, being a Fellow of the Royal Meteorological Society, but was assisted in this by his brother, Sydney Wilson, who took regular meteorological readings at his home in Garstang, which is a fairly central point in the area covered by the book. He also accompanied Wheldon and Wilson on many field trips when the book was being researched. The book is illustrated with photographs taken by Albert Wilson. On publication, the book was very well received and described at the time as "a standard model of what Floras should be" and also as a "credit to the authors and a lasting memorial". A review in *Nature* in January 1908 congratulates the authors for being prepared to sacrifice hours of leisure and recreation for the book's preparation and welcomed their "painstaking efforts". *The Lancashire Naturalist* said of the book "it is one of the most capable botanical works ever compiled". The book included many 'first records', most of which are attributable to Wheldon and/or Wilson, being reported initially in the *Journal of Botany*; however, the authors did not claim them all as original records as they accepted that their study of previous literature was not exhaustive, owing to lack of time.

After the publication of the *Flora*, Wheldon's geographical horizons broadened and he made several visits to the north east of Scotland, many with Albert Wilson. There may have been botanical reasons for this, or possibly it was to do with Sydney Wilson moving to live at Perth prior to 1907. It is probable that Wheldon contributed to other Floras as he was on the committee that initiated work on one for South Lancashire and he probably contributed to Dallman's research for his never published Floras of Flint and Denbigh. He is also credited with contributions to *The Flora of North Yorkshire* (by J G Baker, 1906). It is interesting to wonder whether Wheldon and Wilson were ever contemplating floras of any of the Scottish counties where they spent so much time.

Wheldon contributed to other books, including Green's *Flora of Liverpool*, and to works on Yorkshire and the Isle of Wight among others. One notable work, published in *The Lancashire Naturalist* in 1911, was entitled 'Social groups and adaptive character in the Bryophyta' which Professor McLean Thompson of Liverpool University called a "striking testimony to Mr Wheldon's knowledge and powers of observation". Another notable work was his *Synopsis of European Sphagna*, published by the Moss Exchange Club, which indicates that his interest was not confined to his local areas and that he was becoming a recognised specialist. In fact, his published articles cover topics relating to Scotland, the Isle of Man and Ireland, as well as occasionally to locations outside Britain. Despite his expertise and reputation in botany, he did not lose interest in other areas of natural history. In 1904, he published an article in *The Naturalist* on the nesting habits of rooks, evidence of his lasting interest in ornithology. In all, he published about 100 papers in a wide range of scientific journals over a period of thirty-eight years (see Appendix 1).

Wheldon worked and corresponded with many other botanists during his lifetime. The best known collaboration is, of course, with Albert Wilson (1862-1949), and it was probably Wheldon who introduced Wilson to the study of cryptogams, as he joined the MEC in 1908; like Wheldon, he was also a member of the short-lived Lichen Exchange Club; he continued as an active member of the MEC until after it became the British Bryological Society in 1923, but his involvement was subsequently curtailed by deafness. Wilson had a similar background to Wheldon; they were the same age, his family were in drapery, and he studied pharmacy, qualifying in the same year, 1884. Possibly the two men first met in London while preparing for their examinations. Wilson was a native of Lancashire and lived in Garstang before moving to Bradford and then Ilkely in West Yorkshire, where he was living at the time of the publication of the *Flora*. In later years, he moved to the Conway Valley, North Wales. The two men were born in the same year, 1862, but Wilson enjoyed a much longer life, dying in 1949.

Although the MEC was a relatively small society, two other members lived within half a mile of 60 Hornby Road and it is possible that Wheldon encouraged their interest in botany and bryology. The Rev. Samuel Gasking (1852-1925) who lived in Yew Tree Road, off Rice Lane, was the Anglican chaplain at the nearby Walton Workhouse, later to be Walton Hospital. Gasking joined the MEC in 1899 and was also like Wheldon a founder member of the Liverpool Botanical Society, becoming its first president in 1906. He also appears to have had broad scientific interests as he was also a member of the Liverpool Geological Society. Gasking's botanical interest was cut short by a stroke in 1921 and he died in 1925. Living nearby was another local botanist and long-term friend of Wheldon, William Gladstone Travis (1877-1958), a patent agent's clerk in Liverpool. He joined the MEC in

1910 and was also a member of the Liverpool Botanical Society, becoming president in 1916. Travis, originally from Kirkdale, chaired the LBS committee for the *Flora of South Lancashire*, although he did not live to see it completed. He was a very enthusiastic botanist and also a gifted linguist; he was a very private man who remained unmarried. He was apparently quite dour, but with a very dry sense of humour. His brother Charles was also a member of the LBS and the two men shared a home at 9 Barton Road until Charles moved to Crosby where he died in 1949. The Travis brothers also had an interest in palaeoecology, and, like Wheldon, W. G. Travis was a pioneer of ecology. They collaborated on several articles, most notably *The Lichens of South Lancashire* in 1915.

Wheldon was also a long-term friend and botanical colleague of Arthur Augustine Dallman (1883-1963). It was Dallman who as a young man of about 23 sent a letter to local botanists which resulted in the founding of the LBS in 1906 and he became its first secretary. His father, the Rev. Augustine Dallman was for many years a vicar in Preston, but by 1905 he was living in Prospect Vale near Newsham Park in Liverpool. Arthur was also living there at the time of the foundation of the LBS, having moved to Liverpool in 1901. He studied botany and chemistry at school in Preston and later became a schoolmaster at various local institutions on Merseyside. He was a Fellow of the Chemical Society. After his marriage, he lived on the Wirral in Higher Tranmere, later moving to South Yorkshire to take up a teaching post. He appears to have had a fondness for North Wales as he made many field trips there, regularly taking holidays there and working on Floras for the counties of Flint and Denbigh. Dallman must have first met Wheldon soon after his arrival in Liverpool as he was certainly a frequent visitor to Hornby Road before the formation of the LBS. The Dallman and Wheldon families were friends for many years.

Another local botanist who no doubt would have been a regular visitor to Hornby Road was the Rev. William Wright Mason (1854-1932), jointly vice-president with Wheldon of the LBS in 1906. Mason was a Lincolnshire man, and until 1892 was the vicar of Leverton in that county. He was a member of the Botanical Exchange Club and in 1901 was living in Balliol Road in Bootle. By 1905 he was the vicar of St Mary's in Derby Road, Bootle. He and Wheldon made several botanical expeditions together, for example to Dolphinholme and Abbeystead in 1901, and he is credited in *The Flora of West Lancashire* as a supplier of botanical records.

Later, Wheldon worked with John Wilson Hartley (1866-1939), mainly with regard to the Isle of Man. The two men made trips there together and it is likely that Hartley introduced Wheldon to the island. Hartley was at one time a member of the Liverpool Botanical Society, although he was actually a native and resident of Carnforth, where he kept a shop. It is possible that the two met while Wheldon was researching *The Flora of West Lancashire* and although Hartley is not mentioned as a contributor to the book, there is a record of him collecting at Kellet, near Carnforth with Wheldon and Wilson at some time before 1906.

It is probable that Wheldon knew, knew of or corresponded with most of the leading amateur botanists (and some professional ones) of his time. Evidence from his letters shows that he was in regular contact with Daniel Angell Jones, a prominent North Wales botanist, H. N. Dixon, an acknowledged expert on mosses, as well as Professor Weiss of Manchester University, and Professor McLean Thompson of Liverpool University, who was later to write an obituary of Wheldon in the *Proceedings of the Linnean Society*.

After the publication of *The Flora of West Lancashire*, a committee was formed by the Liverpool Botanical Society to prepare a companion volume for South Lancashire, to be published under Travis's name. Unfortunately the book took longer to prepare than was anticipated and also ran into some difficulties. Although it was almost completed by the early 1920s, the work was not published until 1963. Travis died in 1958, but the book was titled *Travis's Flora of South Lancashire*, being updated and completed by local botanists Savage, Heywood and Gordon. In the foreword to the book, which had been intended as a companion to West Lancashire, the reasons for its late appearance are given mainly as financial but the death of Wheldon and the departure of some other keen members also contributed to the delay. When work started on South Lancashire it was seen by some as an extension of West Lancashire, prompting Travis to write to the *Lancashire Naturalist* pointing out that it was to be a publication in its own right. It is interesting to contrast the success of the work on West Lancashire, produced by two men essentially as a private venture with that on South Lancashire, the work of a committee which took much longer to complete; however, South Lancashire covered a far greater area (1,130 square miles) than West Lancashire (492 square miles), one of the smallest of the vice-counties.

Wheldon was an acknowledged expert in bryology especially in two groups, the *Sphagna* and the *Harpidia*. He would receive parcels of these to examine from other British botanists, as well as from Europe and North America. According to Professor McLean Thompson writing after Wheldon's death, his knowledge of European species was often called upon by European botanists. His own collection was extensive and contained most British species and some from abroad. He was also an expert on lichens, but one area he did not study very closely was fungi. However, his elder son Harold shared his father's interests in botany and he took on the fungi, both of them working together on this group.

Wheldon spent most of his leisure time working on botany, particularly bryology. Probably the whole of his twenty-one days' annual holiday must have been spent on botanical research; not much time would have been left for family holidays. Professor McLean Thompson later wrote of Wheldon that "he seized every opportunity to journey widely in the British Isles to see rare plants in the field or to confirm and extend earlier observations". He would walk for miles, sometimes twenty or thirty miles per day on country lanes, fells and sand dunes, collecting specimens, and then when he got them home he would study, classify and catalogue them. In the process, he built up a herbarium which the *Lancashire and Cheshire Naturalist* described as "probably the most complete". His obituaries mention that he had a pleasant nature (a fact confirmed by his daughter-in-law Madge who was very fond of him) and was always willing to help and encourage anyone with a genuine interest. At the same time he was very professional in his approach and insisted that others should adopt the same approach to their botany. His home was always open to those who shared his interests or sought help and guidance; one local botanist described 60 Hornby Road as the "Mecca" for all botanists in the area. In addition, Wheldon lectured from time to time to the Liverpool Botanical Society and other similar organisations using lantern slides which he had prepared himself, and usually supplied any meetings he attended with interesting specimens. He also acted as a specialist for the *Lancashire and Cheshire Naturalist* from 1908.

In 1922, Wheldon was awarded an Honorary MSc by the University of Liverpool, an award about which he was very modest. This was a notable achievement in that it was for work conducted entirely in his spare time, and not in connection with his profession. At the ceremony he was presented with his degree by Lord Derby and Professor Daly, the Dean of the Faculty of Science, said that "James

Alfred Wheldon for many years has found time amidst his official duties to prosecute research in botany and on the Flora of Lancashire. As a leading authority on mosses his work stands in high repute among the botanists of today". Wheldon was one of only five people to receive Honorary Masters degrees from the University in that year, but among those who received doctorates in that year were two people who had made considerable contributions to Liverpool: Francis James Chavasse, second Bishop of Liverpool and responsible for the building of the Anglican Cathedral; and J. A. F Aspinall, engineer and later general manager of the Lancashire and Yorkshire Railway, whose pioneering electric trains ran through the cutting adjacent to 60 Hornby Road. After his retirement in 1921 from Walton (as the senior pharmacist in the Prison Service), it was Wheldon's intention to re-organise the herbarium at Liverpool University, something he looked forward to, but this task was never completed due to illness.

For thirty years Wheldon's place of employment was the prison in Hornby Road, Walton. It was built between 1850 and 1854 on a radial principle and originally had accommodation for 1,000 prisoners of both sexes. (In 2009 there were 1,184 inmates, all male.) Walton replaced Kirkdale Gaol which closed in 1892 although by that time most of the prisoners had already been transferred. Wheldon would have been at Walton when the last of these transfers took place. After Kirkdale closed, all local executions took place at Walton. By the abolition of the death penalty in 1964, sixty-two prisoners had been hanged there, including two women. On average there was one execution per year, but not every year; for example, there were none in 1905 or 1906. For most of Wheldon's service there, executions would be a regular, if only occasional event. The infamous Pierrepoint family were regular executioners at Walton; during Wheldon's time Henry Pierrepoint was involved and in later years

Albert officiated. Other 'hangmen' included John Ellis who carried out fourteen hangings in the years between 1910 and 1923, often working with one of the Pierrepoints as at, for example, the hanging of Thomas Seymour in 1911. The tolling of the prison bell before an execution would be well known to the residents of Hornby Road, well remembered by Wheldon's grandson Jim, as would the crowds that often gathered in front of the prison at the time of an execution. Viewed from the 21st century this seems extraordinary, but at the time was seen by most people as a necessary part of the judicial process.

The harshness and brutality of some aspects of Wheldon's working environment in contrast with the more gentle pursuit of botany are quite marked. During this time some suffragettes were on remand at Walton and embarked on a hunger strike. Controversially, they were force fed – Lesley Hall and Selina Martin in 1909 and Lady Constance Lytton and Rose Howey in 1910. This was the background to Wheldon's professional life. One of his great friends and colleagues Dr Arthur Price was a medical officer at the prison, often confirming death after an execution (as in the case of Seymour mentioned above) and also apparently being involved in the force feeding of the suffragette prisoners. Another aspect of Price's work is illustrated by the Seymour case. During his time in prison, Seymour was closely observed by Dr Price in order to look for signs of insanity. None were found and he was hanged. Wheldon must have worked closely with medical staff at the prison, and it is on record that he sometimes carried out duties associated more with a medic than a pharmacist.

An integral part of the job at the prison was the house at 60 Hornby Road. Described as a 'house with six rooms', it was separated from the prison only by the railway cutting, and would be very convenient for his work. However, on retirement, the Wheldons had to seek alternative accommodation. After a long search, a new home was found, and they moved only a short distance away to the relatively new Marchfield Road, Orrell Park, and a pleasant terraced property with a rear garden overlooking a small park known as Devonfield Gardens. It was here that James Alfred Wheldon died on Friday 28 November 1924, working on his botany almost to the last. He had just completed an article for the *Journal of Botany* on *Additions to the Scottish Sphagna* and had prepared samples for distribution to the BBS. Some of his unfinished work was to be completed by his son Harold, but in fact it was completed by others as Harold lost interest in botany after the First World War. Wheldon was only sixty-two when he died and had only had a short period of retirement from the Prison Service. He doubtless would have done much further botanical work had he lived longer, but in his last year he suffered with prostate cancer which necessitated two spells in hospital and which eventually killed him. At the time of his death, Wheldon was sufficiently well known locally for the *Liverpool Echo* to report it on 1 December. In addition to the family announcement in the classified section, the paper noted that "the death has occurred at his residence in Aintree of Mr James Alfred Wheldon, for thirty years pharmacist at Walton prison. He was widely known in international botanic circles and on his retirement three years ago was the senior pharmacist in the prison service. He was decorated with the Imperial Service Medal". The *Liverpool Daily Post* had reported his death on Saturday 25 November stating that it occurred "after much suffering, patiently borne". James Alfred Wheldon was buried in Walton Cemetery on 3rd December 1924, the service taking place at the cemetery chapel; he was buried with his wife Catherine and brother-in-law Septimus Simpson. Probate was granted jointly to Harold and William on 31 December. He left £1,115. 9s. 9d.

The funeral was reported in the *Liverpool Daily Post & Mercury* on Wednesday 3rd December.

"The funeral of the late Mr J. A. Wheldon, who was an officer on the staff of His Majesty's Prison Walton for thirty years took place yesterday at Walton Cemetery, The Rev. Osbourne Gregory officiating.

A large detachment of officers from His Majesty's Prison marched in procession before the hearse from the residence of the deceased and six acted as bearers. A large number of retired staff of the prison formed a double line at the entrance to the cemetery. In addition to a great number of wreaths from relatives and friends, there were beautiful floral tributes from the medical and nursing staffs at the prison."

The paper went on to list the mourners at the funeral. Harold and William lead the mourners together with Mrs Wheldon (assumed to be Madge) and Harold's eldest son, Basil. Others present included Rev. Gasking and W. G. Travis as well as Major and Mrs Price. Dr W. A. Lee and the Misses Bangham and Warhurst were there to represent the LBS.

CHAPTER 3

BOTANICAL JOURNEYS

During his lifetime, Wheldon made extensive botanical journeys to study, record and collect specimens. From existing records it appears that some years were busier than others, but he would also have spent a great deal of time at home, working on his herbarium, writing articles and papers, and engaging in the business of the LBS and MEC. It is recorded in obituaries that he used virtually all of this annual leave from the prison on his botany and it is interesting to look at where he visited at various times in his life. These trips would be in addition to purely local ones near to his home, such as to the coast between Liverpool and Southport.

Records of his journeys in the *Flora* show relatively few prior to 1898, although he made collections in the Yorkshire area from 1876 to 1891. However, 1898 and 1899 were very busy years, possibly when much of the fieldwork for the *Flora* was undertaken, most of his trips being to locations in the West Lancashire area. He made regular visits to Preston and to nearby Longridge and Longridge Fell, and also to the Lancashire coast, to Lytham St Annes, Blackpool, Fleetwood, Morecambe, Heysham, Lancaster and Silverdale; many of those destinations were visited again in 1900, with the addition of the docks at both Fleetwood and Preston and many other places in the West Lancashire area. From 1901, records show that his journeys became less frequent, although destinations in Lancashire, such as Fleetwood and Lytham St Annes, were still visited regularly, and from this time Wheldon's writings become more frequent, suggesting that he might have devoted more time in these years to writing and research. After the publication of the *Flora*, annual trips were made to Scotland, particularly to the north-east, as well as more local visits to North Wales, the Lake District and the Isle of Man.

The type of location Wheldon visited in his search for plants is also interesting; he was not always wandering the moors, mountains and sand dunes. The *Flora* records finds made at Carnforth railway station, a canal at Garstang, a reservoir at Grimsgagh and even waste ground at Preston Docks. A study of the records of Wheldon's collections held at the National Museum of Wales also gives an interesting insight into his activities. As would be expected, most of his collecting activities prior to 1890 were in Yorkshire and as some of those samples still exist it would appear they might have escaped the York fire. After the fire the Wheldons lived briefly in Ashton-under-Lyne where he was soon busy in the area, with collections being recorded in Greenfield, Hollingworth and Oldham, all within easy distance of Ashton-under-Lyne. In 1891 he made his first visit to the Lancashire coast at Southport and also recorded collections close to his new home at Fazakerley and Walton. During the next two years, most of his collecting appears to be done locally on the South Lancashire coast and on the Wirral, although trips to Llandudno and the Great Orme seem to have taken place in 1892 and 1894. The years 1894 and 1895 also saw trips back to Yorkshire (Teesdale and Ingleborough). A similar pattern was followed in 1896 and 1897, mainly local to Liverpool, but with some trips to Yorkshire, possibly to visit family. From 1898 the work on the *Flora* started in earnest and the majority of Wheldon's collections in the years up to 1904 were from the West Lancashire area. Presumably the next couple of years were spent preparing the *Flora* for publication, as visits to that area became much less frequent.

From 1905, botanical journeys became much longer and more ambitious and involved trips to Scotland, the Isle of Man and the Lake District. This is possibly because by this time Wheldon's interest was concentrated on the *Sphagna*, and the wetter conditions of those areas were and are better for their study. Indeed it is possible that after about this time most of his botany involved lichens and mosses, as he admitted in a letter to Arthur Dallman in 1924 that he needed to "refresh his memory on flowering plants". Visits to the north-east of Scotland began in July 1905 with a visit to Perthshire and Easterness. Wilson was also present and it is possible that the trip was combined with a family holiday as evidence shows that the 17-year-old Harold was there as well. As mentioned above, the fact that Sydney Wilson was now resident in Perth may have had an influence on the choice of location.

1905 and 1906 appear fairly quite years (apart from writing up the *Flora*), but they include Wheldon's first recorded trips with Dallman to Llyn Helyg and Mostyn in North Wales. However, another visit to Scotland in July 1908 was the first of six consecutive annual visits north of the border. Again the destination was Perthshire and Easterness, and again Wilson was also present. Earlier in that year the two men visited Westmorland and in the autumn a trip was made to the Isle of Man with J. W. Hartley. 1909 followed a similar pattern – Westmorland in May, Banffshire and Easterness in July, and the Isle of Man in September; once again Wilson and Hartley were the companions. Another Scottish trip to Perthshire and Argyll was undertaken in Jul 1910, and Perthshire was again the destination in the summers of 1911, 1912 and 1913. An interesting venue in July 1913 was Ryton on Dunsmore in Warwickshire, where Harold lived, and there is a record of a further collection there in September 1915. Significantly 1913 and 1915 were the years in which Wheldon's grandsons, Basil and Roland, were born, so it is likely that a family visit was combined with some botany.

In June 1914, on the eve of the First World War, Wheldon again visited the Isle of Man with J. W. Hartley, but very few other collections are recorded in that year, or 1915, possibly as a result of the war and possibly due to Catherine's death. The year 1916 was more productive and visits were made to Cumberland and Westmorland with Wilson in June, and to the Isle of Man in August, interestingly with Harold, who must have had some leave from the Ministry of Munitions. Possibly they had a holiday together to help get over Catherine's death eleven months previously. During 1917 there were few long journeys, although a trip may have been made to the Isle of Man with Doris, and collections were generally made in the local area in the company of W. G. Travis and D. A. Jones. Further local collections, mainly with Travis, are listed in 1918, together with visits to Westmorland and the Isle of Man in June of that year. In August 1919 a visit was made to Llanberis and Snowdon with D. A. Jones.

In August 1920, Wheldon is recorded as making botanical collections on Dartmoor. It is possible that he was with Harold, as it is known that Harold was fond of visiting Devonshire to paint – so maybe they were enjoying some time together. September of the same year saw another Isle of Man trip and collections were made with Hartley. Wheldon's last visits to the island were made in September 1922, again with J W Hartley and in September 1923 with his old friend Albert Wilson. In 1923 a visit was also made to Westmorland, but by 1924 Wheldon was terminally ill and his last recorded collections, between spells in hospital, were local ones at Rainford and at New Ferry with Travis.

Today all the places mentioned could be easy reached quickly by car, but at the time the only means of travel would be by railway, as reliable motor transport was not really available until after the First World War. From the railhead, the journey would be either by foot or some form of horse transport; this would have added somewhat to the time required for the journeys, although time in the train could have been used for reading and writing up notes. The diaries of Arthur Dallman give an interesting insight into the ways in which botanists travelled in those days. The train was the obvious mode of transport and several of Dallman's excursions to North Wales start at Woodside station in Birkenhead, although he does refer on some occasions to cycling; for example, he cycled from Birkenhead to Flintshire in August 1912, and it is likely that cycles would have been taken on the train. Whether or not J. A. Wheldon used a bicycle is unknown.

The railway would be the only choice of transport for Wheldon's botanical journeys. Compared to today, the journeys overall would have been slower. Although there were some speedy services, for example to Manchester, they were less frequent, but the range of available destinations would have been greater and the use of through coaches meant that fewer changes of train might be needed. For most of his intended destinations, Wheldon was very well provided with convenient trains. From his home in Hornby Road, a local train to Liverpool Exchange Station from either Walton Junction or Preston Road would take less than ten minutes, and from Liverpool Exchange, the Lancashire and Yorkshire Railway ran services to all parts of Lancashire, to Scotland and to York and the North East. Perhaps the availability of these services influenced Wheldon's choice of destinations.

During the years 1898 to 1906, Wheldon made regular journeys into West Lancashire in connection with the *Flora*. A look at the railway map of the time shows that the west of the vice-county was very well supplied with railways, with the main line from Preston to Kirkham affording three different routes into Blackpool, one of which went via the coast giving access to St Annes and Lytham, both very regular destinations, and another giving access via a branch to Fleetwood and Wyre Dock, again both visited many times by Wheldon. These journeys from Liverpool would be very straightforward. A branch line also ran from Garstang to Knott End; again both destinations mentioned in the *Flora*. The east of the vice-county was a different matter: the main west coast line almost exactly marks the boundary between the east and west sections and only one line penetrated into the much more difficult country to the east, namely a branch from Preston to Longridge with an intermediate station at Grimsargh. The rest of the eastern area would have to be explored on foot; some destinations such as Mallowdale Fell, for example, would involve quite a journey.

It is interesting to try to imagine these journeys. As an example, in August 1901 Wheldon went to the area around Dolphinholme and Abbeystead in the company of William Wright Mason. This would have involved a train journey to Bay Horse, a small station between Preston and Lancaster which enjoyed only a very sparse train service and would have involved a change at Preston; from there a walk of about two miles uphill would have taken them to Dolphinholme and then a further four miles on to Abbeystead, stopping frequently to examine specimens and no doubt deviating off the road many times, before returning to Bay Horse for the train. A missed train would involve a very long wait at the station. This journey, however, would be comparatively easy compared to the journey to Mallowdale Fell, mentioned above, which would have been quite an expedition, the destination being at least ten miles from the railway and about 900 feet above it.

Liverpool was, and is, the departure point for sailings to the Isle of Man, so reaching that island was relatively simple. Internal transport would be provided by the Isle of Man and Manx Electric railways, and by the time Wheldon was a regular visitor to the island motor transport would have been more reliable and available. Scotland was a much more challenging destination, with many of the locations visited involving complex journeys by train and probably foot. Wheldon's regular visits to Scotland started in around 1905 and were annual for several years after that. They illustrate that he must have been a very fit man. He was well into his forties by the time of his first visit and they involved long walks over difficult terrain and considerable climbs, often in less than perfect weather, and overnight stays in mountain bothies. In addition, a contemporary photograph shows Wheldon botanising in tweed suit and Homberg hat; none of today's lightweight waterproofs were available to him then. He tackled these trips with enthusiasm and his writings suggest that he was very keen on the area. His general level of fitness and obvious love of the outdoors must have made his later incapacity due to cancer much more difficult to bear.

A flavour of the nature of the Scottish expeditions can be found in the written accounts of the visits and these suggest that they were not easy rambles in the hills. Some considerable climbing was required over what must have been fairly rough ground, and despite being in high summer, they did not seem to have much luck with the weather. In July 1908 Wheldon was in Inverness-shire with Wilson. In the introduction to their paper "Inverness-shire Cryptogams" the authors give some detail of the nature of their botanical expeditions.

> "In July this year we spent a few days in East Inverness-shire, making Aviemore our

headquarters. Our original intention was to examine some of the principal summits of the Cairngorm range, but this was to a large extent frustrated by the extremely unfavourable weather conditions which prevailed during our visit. Fortunately the great Forest of Rothiemurchus provided an attractive alternative to the mountain peaks, and we gave it more attention than would have been the case had the mist and rain permitted us to spend more time on the mountains.

This extensive forest tract has an area of nearly one hundred square miles, and we traversed it in several directions by intricate paths between Aviemore and the foothills of Ben Macdhui and Braeriach which abut on the Larig Ghru and Glen Eunach."

"Inverness-shire Cryptogams" by Wilson and Wheldon. *Journal of Botany* 1908

The following year they returned to the Cairngorm Mountains for four days in July, and reported that they had found a number of mosses and lichens additional to those found on their previous visit. This year however they extended their searches into Banffshire, to the "most alpine" portion of that county at the head of Glen Avon. The journey is then described.

"Briefly stated our route was as follows: Starting from Kincraig station on the Highland Railway, we ascended Geal Charn and Sgoran Dubh from the Glen Feshie side, descending into Glen Eunach by a corrie at the head of the glen. After spending the night at the upper bothy, we ascended Braeriach as far as the tarn called Loch Coire an Lochan, situated at an altitude of 3,250ft – the highest sheet of water of its size in Britain – on the north side of the mountain. Here we worked a small portion of ground by some patches of snow, at the base of granite cliffs, which rise above the loch, and found several mosses of considerable interest; but a detailed examination of the rocks or further progress up the mountain was impossible owing to mist and the fury of the wind."

"Inverness and Banff Cryptogams" by Wheldon and Wilson. *Journal of Botany* 1910

This poor weather lead to their descent back to Aviemore and the following days were spent on the banks of the Spey at Craig Ellachie and on a climb of Cairngorm and Ben Macdhui. Their search for high alpine species was cut short by "a good deal of snow" still lying at the higher levels (in July).

(Kincraig station, now closed, was situated between Kingussie and Aviemore, close to Loch Insh. Cairngorm has an altitude of 4,085ft, Ben Macdhui is 4,295ft, and is second only to Ben Nevis as the highest peak in Britain)

This was not apparently the first time that they had attempted to climb Braeraich. In August of 1909, Wheldon received a letter from William Ingham, a botanical friend from York, which notes that he was sorry that they had had "such bad luck again with Braeriach".

In the summer of 1910 Wheldon and Wilson's destination was Perthshire and Argyll, their journey following part of the famous West Highland Railway. They are on record as visiting Rannoch Moor, Loch Rannoch and three further Munroes: Ben Challum (3,363 ft), Ben Lui (3,707) and Ben Dothaidh (3,294) although it is not clear as to whether they actually reached the summits of those mountains.

In 1912 Wheldon and Wilson were in Perthshire and on 6 May that year they ascended Ben-y-Gloe to a height of 3,505ft but once again were frustrated by thick mist.

Many trips would obviously involve overnight or longer stays; for example, records of his visit to the Llanberis area suggest collections were made on several days and obviously the Scottish or Isle of Man visits would involve staying in the area. The costs of these journeys cannot have been insignificant. A description of one such trip is recorded in the diaries of Arthur Dallman. In August 1906, Dallman, enjoying the long holidays which were part of his profession as a schoolmaster, went to North Wales. He journeyed from Liverpool by ferry to Eastham and the cycled from there to Nannerch on the Mold to Denbigh road. He had arranged accommodation in nearby Ysceifiog (for ten shillings a week, bed breakfast and supper) and spent the week exploring the area. At the weekend (25 and 26 August), he was joined by Wheldon, presumably journeying by train (again it would be interesting to know if he had a bicycle), and the two men spent the Saturday botanising at Rhydymwyn, Cilain and Moel Arthur and the Sunday at Llyn Helyg where they made some botanical collections, records of which are still in the National Museum of Wales. Wheldon also took advantage of his botanical journeys to indulge his interest in ornithology and in wildlife in general. The following short articles appeared in the *Lancashire Naturalist* in September 1921. The South Cliff referred to is at Scarborough, where he was living and working at the time of the observation.

BIRD NOTES

"To the instances in which birds succeed in rearing their young by sheer effrontery, I can add one which occurred this year. On the Orrell Park allotment gardens a skylark made a nest at the foot of a clump of "twitch grass" (Agropyron repens), growing as a weed on the plot next to mine. It contained young when I discovered it, and the garden was being dug over a few yards each evening.

When the digger arrived within a short distance of the nest, I intended to warn him. However, before he reached it, the young disappeared, so I did not mention it, and concluded that the nest had been robbed by boys or cats. To my satisfaction, however I found that the birds had flown, and for several days had the pleasure of watching the young larks fluttering about the gardens. In the eighties I saw the nest of a Grasshopper Warbler within three feet of a footpath on the South Cliff, which was used by thousands of people – especially on gala nights on the spa; and I have seen the Snipe nesting within five yards of a high road in West Lancashire."

NATTERJACKS

"Both this year and last must be classed as dull and wet seasons, but one would not expect that to diminish the number of Natterjacks. I saw about the usual number at Ainsdale last year. This year, so far as it has gone, they are, I think, excessively numerous. In one "slack" the ground was alive with them, so that it was impossible to walk without killing several at each step, over large areas. They were scampering in every direction, running quickly, not crawling like the common toad, and no larger than a marrowfat pea, some smaller. This was on July 12th. A very small proportion of these hoards are, however, destined to reach maturity, but in this particular locality they stand a better chance than in the more remote parts of the dunes, as there are usually numbers of people near; and houses in the vicinity. This no doubt keeps at a respectful distance predatory birds which might find the little Natterjacks a dainty morsel."

Harold James Wheldon

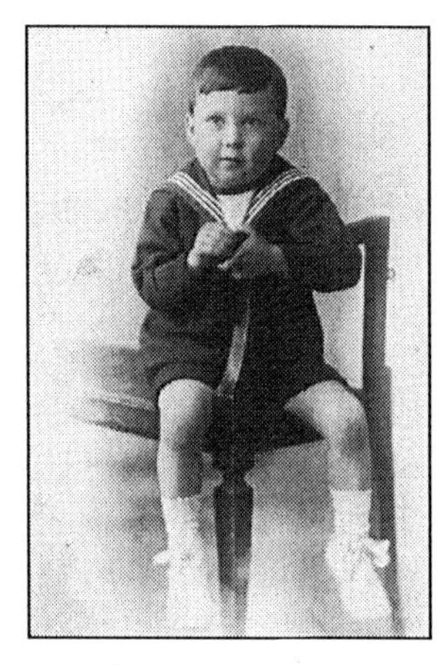

Jim Aged 5 years

James Alfred Wheldon
about 1921

Catherine Wheldon

J A Wheldon MSc

J A Wheldon 1909

Marjorie Howard in 1914

William Alfred Wheldon

26 Marchfield Road

Walton Prison about 1910

60 Hornby Road, Side view

60 Hornby Road

Brook Road Church

Grave at Walton

Imperial Service Medal

PROGRAMME.

7.0-7.30. Reception by the President, A. A. DALLMAN, Esq., F.C.S., and the Hon. Secretaries.

Exhibits.

7.30. Address by the President.

7.40. Concert.

8.15-10.0. Refreshments, Competitions, and Conversazione.

PROGRAMME OF CONCERT.

Violin Solo ... ... "Allegro Brilliante" *(Willem ten Have)*
Miss OLIVE BANGHAM.

Song... ... "The Pipes of Pan are Calling"
Miss OLIVE PARSONAGE.

Song... ... ... "Pals" *(W. H. Squire)*
Mr. EASTLAND STAVELY.

Recitation ... "Miss Withers and the Wasp" *(Jessie Pope)*
Miss MURIEL SLADE.

Song... ... "I Think" *(Guy d'Hardelot)*
Miss CONNIE OGLEY.

Accompanist:
Miss GLADYS PARSONAGE.

A typical LBS Soiree

SYLLABUS.

1906.

Sept. 10—"A Botanical Ramble on the Lancashire Fells." (*Illustrated*) **Mr. J. A. Wheldon, F.L.S.**

Oct. 10—"Notes on the Flora of Snowdonia." (*Illustrated*) ... **Mr. W. Travis.**

Nov. 12—Microscopical Evening ... ...

Dec. 10—"The Life History of an Apple." (*Lantern Lecture*) **Rev. T. J. Walshe, F.R.A.S.**

Dec. 19—"Plants of the Fenland." (*Lantern Lecture*) **Prof. R. H. Yapp, M.A.**

1907.

Jan. [illegible] —**Annual Meeting.**
"The Alien Flora of the Liverpool District"

Jan. 14—*Title to be announced later.* **Rev. S. Gasking, F.L.S., B.A.**
Held over to March owing to date of Ann. Meeting

Jan. 28—"The Scots Fir." (*Lantern Lecture*) **Mr. W. T. Haydon.**

Feb. 11—"Botanical Ideals.—Yesterday and To-day." ... **Miss E. E. Appleton.**

Feb. 19—"The Flora of South Africa." (*Lantern Lecture*) **Prof. F. E. Weiss, D.Sc., F.L.S.**

Mar. 11—"The Relation of Structure to Function." ... **Miss H. Roberts.**

Mar. 27—"Plant Hairs." (*Lantern Lecture*) **Mr. W. R. Croston.**

April 8—"The Quillwort." (*Lantern Lecture*) **Mr. A. A. Dallman, F.C.S.**

April —Demonstration Evening.... ... ...
Several short papers illustrated by demonstrations, dealing with the methods of study, collection and preservation of
Flowering Plants. Mosses.
Seaweeds and Algae. Lichens. Fungi.

One or two Winter Field Meetings will probably be arranged for the study of Cryptogamia, Trees, &c.

LBS programme 1906

CHAPTER 4

FAMILY LIFE

James Alfred Wheldon married Catherine (also known as Caroline) Simpson on 17 February 1887 at All Saints Church, Northallerton. She was born on 8 November 1865 at the Royal Oak Hotel in Bedale, North Yorkshire, where her father William was the licensee. She was the youngest of ten children and continued to live at the Royal Oak after the death of her parents, when brother John took over the hotel. In 1881 she was described as a waiting maid at the Royal Oak and was living there with John and her youngest brother Septimus Herbert, only fourteen months her senior, with whom she appears to have had a close relationship. Indeed they died within six months of each other in 1915 and share a grave with James Alfred Wheldon in Walton Cemetery. By the time of her marriage she had moved from Bedale to Northallerton, possibly with Septimus. Certainly by 1890 the Simpsons no longer kept the Royal Oak in Bedale and by that date Septimus was married.

James Alfred and Catherine had three children: the eldest, Harold James, was born in York at 20 High Ousegate on 7 July 1888 and their second son, William Alfred, was born on 4 September 1889, also at 20 High Ousegate. After the fire in York on 31 December 1890, the family lived briefly at 32 Langham Street, Ashton-under-Lyne, where Doris Mary was born on 18 July 1891. During their stay in Ashton-under-Lyne, Wheldon was employed as a commercial traveller, probably for a chemist or druggist, but he soon obtained the post at Walton Prison where he remained until his retirement in 1921/1922.

On moving to Liverpool, the family's first home was at 9 Chelsea Road, Walton Vale; this was, and is, a short cul de sac of terraced houses about fifteen minutes' walk from the prison. After a few years the family moved to 60 Hornby Road, Walton, a prison property. It was part of a terrace of houses, all occupied by prison staff, which ran from the corner of Hornby Road and Rice Lane as far as the Lancashire and Yorkshire Railway line to Ormskirk and Preston. On the other side of the railway stood Walton Prison, built in the 1850s. Number 60 Hornby Road was the last of the terraced houses, overlooking the railway cutting, and at the time that the Wheldons moved in, the houses stood in some isolation compared to today, and were all occupied by prison staff. There were no buildings on the other (south) side of Hornby Road, apart from a mission room opposite the prison. The entrance to Walton Junction Station was almost opposite number 60 and further down Hornby Road towards Rice Lane was the entrance to Preston Road Station (now Rice Lane) on the line to Wigan and Manchester. Today the area is built up as far as Bootle, but when the Wheldons first moved to Hornby Road it was a dead end and there were fields between them and the rows of terraced houses slowly expanding up the hill from Bootle and the docks. A short distance behind the house ran the Midland Railway branch, giving access to those docks; the sound of the steam engines pounding up the hill would be a regular feature, well remembered by Madge and Jim in later years. 60 Hornby Road no longer exists, although most of the other prison houses still stand. The site of number 60 is now a car park for the prison.

The area was very well provided with transport. Frequent trams ran down Rice Lane from Aintree and Fazakerley to Liverpool and both stations mentioned provided services to Exchange Station in Liverpool. In addition, Wheldon might have been able to start his journeys to the north from Walton Junction to Ormskirk, changing there for trains to Preston and beyond.

As far as the education of his children was concerned, no record has been found of Harold's education, but it is likely that he had a similar education to his younger brother William, who attended the Liverpool School Board establishment at Northcote Road. This was a large school of over 1,000 pupils of all ages, so it is probable that William started here as an infant. The school was situated some distance down Rice Lane towards Walton on the Hill and William was a pupil there until 1901. After 1905, a new school was opened on Rice Lane, just around the corner from Hornby Road, which would have been much more convenient. In 1901, he moved to the Oakes Institute on Rice Lane, almost opposite the junction with Hornby Road. This establishment, a private school, was run by W. A. Oakes. It offered qualifications validated by the College of Preceptors, an organisation set up in 1849 to regulate the qualifications offered (and the teachers) at private schools. William remained there until at least 1903. No record has been found of Doris's education. The Wheldons do not appear to have been members of the Anglican Church, their parish church being St John the Evangelist, Walton, very close to Hornby Road, but attended the Brook Road Wesleyan Methodist Church, Rice Lane. It is possible that Catherine's family, the Simpsons, were Anglicans, as she was baptised in her parish church, but the Wheldon family were certainly Wesleyan Methodists. James Wheldon Junior had been a member of the Wesleyan Methodist Church in Bedale and Northallerton; Robert Wheldon, his brother, is also known to have belonged to the Wesleyan Church, and, as mentioned above, James Alfred and all his siblings had been christened in the Wesleyan Chapel in Northallerton. It is not known how regularly they attended Brook Road, but both sons were married there, Catherine's memorial service was held there and grandson James Alfred (Jim) was baptised there. It is not certain whether James Alfred Wheldon was a freemason, but his father was (James Wheldon Junior was inducted into the Northallerton Anchor Lodge in 1875), as was his uncle George, and son William was for most of his adult life.

By 1911, all three children had left school and were in employment. Harold was a draper's clerk (which is interesting in view of the family business in Yorkshire), but the term clerk probably understates Harold's role as at this time he would certainly have been studying for accountancy examinations. William was employed as clerk to a provision merchant, and Doris was a shorthand typist. They were all still living at 60 Hornby Road, which must have been quite crowded with five adults and a very large botanical collection to house.

In 1915, a momentous year for the Wheldons, Harold's second son Roland was born in Warwickshire on 15 March and on the same day Septimus, Catherine's brother, died in Liverpool. On 26 September tragedy struck when Catherine died at home suddenly from a haemorrhage, aged only fifty. She was buried in Walton Cemetery along with her brother. She left an estate of £128.1s.3d. Her death must have been a huge blow to Wheldon as they had been married for 28 years and she had enthusiastically supported his botanical activities. She seems to have been a gregarious, cheerful, outgoing character and would have been greatly missed.

Obituary Catherine Wheldon

(*Proceedings Liverpool Botanical Society* 1915)

> "One of the most beloved members of the Liverpool Botanical Society, the wife of a former president and an enthusiastic worker, Mrs J. A. Wheldon, died very suddenly on the morning of Sunday 26th September. She had been in her usual apparent health and her end was quite unforeseen. Her tragic passing was the occasion of a manifestation of widespread and genuine regret among the membership as a whole who had found her to be the spirit of sociability to a marked degree. The early days and in fact the inception of the society evoked her warmest sympathy and although she never appeared to be a technical expert, she took a keen interest in natural science. The deep sympathy of the members was evidenced by the large attendance at the memorial service which was conducted in Brook Road Wesleyan Church and by the unique floral tribute consisting of a harp-shaped token, in which heather and other appropriate wild plants found a prominent place."

As a memorial to Catherine, the Liverpool Botanical Society members subscribed for an oak lectern, described at the January meeting in 1918 as follows:

> "An oak lectern, provided with reflector plug and lamp holder, subscribed for by a number of friends as a memorial to the late Mrs Wheldon, was formally presented by Dr Lee (the Society treasurer) who spoke of Mrs Wheldon's delightful personality. The gift was accepted by the president (Dallman) on behalf of the Society. The secretary and Mr Dallman spoke of the life and influence of Mrs Wheldon. It was in her drawing room that the

> idea of the Liverpool Botanical Society was first discussed and from that time Mrs Wheldon took an active interest in its welfare and had done a great deal to promote its success."

A plate on the lectern bore the following inscription:

> "In memory of Mrs J. A. Wheldon, whose life and work were highly esteemed by friends associated with her in the Liverpool Botanical Society."
>
> "Let her works praise her in the gates"
> Proverbs XXXI.31

Her brother Septimus Herbert, known in the family as Uncle Septimus, had followed the family trade and in 1890 was an innkeeper at the Royal Victoria Hotel, Ormesby in North Yorkshire and later at The Sutton's Arms, Elton in County Durham. He married Hannah Nelson whose father had also been an innkeeper at the Nag's Head in Northallerton. In 1901, Septimus was in Elton with Hannah and her mother Christina, but in 1907 Hannah died and apparently Septimus moved to be nearer his sister. In 1911, he was employed as a barman and cellar man in New Brighton, but by 1915 had moved to Liverpool and lived at 4 Parkinson Road, Walton, a short distance from Hornby Road and his sister Catherine, and worked locally as a barman. In March 1915, he was admitted to the David Lewis Northern Hospital in Great Howard Street, Liverpool, where he died aged fifty.

Catherine's sudden death must have been a major loss for Wheldon. While claiming no expertise, she had supported him in his interest in botany, probably at the expense of having few, if any, holidays, and had played a major role in the LBS. After her death he continued with his work at the prison and with his botanical work. At this time, only Doris was at home, Harold was married and living in Warwickshire, and William was working abroad in Sierra Leone, West Africa. Doris took on the role of looking after the house and her father. It is likely that he was also supported by Marjorie Howard (Madge), his future daughter-in-law, who married William (Bill) in 1919. Madge was born in Birkdale on 29 June 1893, the daughter of Nathan Howard, a bricklayer. She lived in Birkdale until 1902, but her movements for the next ten years are unknown and Madge remains an enigma. She became known to the Wheldon family in about 1912 and before her marriage in 1919 she was working as a clerk at a firm of cotton brokers at the Cotton Exchange in Bixteth Street, Liverpool (although she had wanted to train as a nurse) and was living at 40 Chatsworth Avenue, Orrell Park. Madge had known Bill since at least 1912. She remembered that they had been out together on the sands at Seaforth on the day the 'Titanic' was lost; they found out about the tragedy in a special edition of the *Liverpool Echo* bought on their way back from Seaforth. It is not known how Madge and Bill met, but it may have been through Brook Road Church, where Madge had been a member of the choir for some years.

For the last few years of his life Wheldon lived with William and Madge and grandson James Alfred (Jim or Jimmy) first at 60 Hornby Road and then at 26 Marchfield Road. He took special pride in Jim as he was the first of his five grandchildren to be born in the area (Jim was born in Nurse Tyson's Nursing Home in [appropriately] Moss Lane, Orrell Park on 21 April 1920 and named after his grandfather at the insistence of Madge), and in his retirement he spent a lot of time with him. Jim had a few early memories of his grandfather: his earliest memory, when aged about 4, was a visit with his grandfather to 'Mr Travis' (Wheldon's botanical friend W. G. Travis); he also remembered that Travis had an enormous cat and only in later life did he realise that the cat was quite normal

– he was very small! He also remembers his grandfather taking him to some railway sidings, later identified as Aintree Sorting Sidings at Ford, where there were some stored locomotives. Wheldon and Jim climbed onto one of the engines and lit some old newspapers in the firebox to simulate the engine working.

The Wheldon family had many friends in the area. Apart from the fellow botanists, such as Travis, Gasking, Wright Mason and Dallman, all their neighbours in Hornby Road would be known to them as fellow prison employees. One particularly close friendship was with the Price family. Arthur Price (1854-1933) had been a major in the Royal Army Medical Corps and was for many years medical officer at Walton. He did not live in prison accommodation, but resided nearby in Moss Lane with his wife Ada. No doubt the two men would have worked closely together at the prison, but were great friends socially as well. Arthur Price died in October 1933 and Ada lived on until December 1936. It is evidence of the friendship between the two families that she spent the last months of her life being looked after by Madge and Bill, and at the time of her death was living with them at 10 Cranmore Avenue, Crosby. To this day the Wheldon family has books, jewellery and an 18th century grandfather clock given to them by the Prices, who had had no children. Another friendship was formed with Arthur Downer, a clerk at the prison who latterly lived at 58 Hornby Road. Mrs Downer was godmother to Jim.

There was also another member of the Wheldon family living in the area, although whether they were in contact, or even knew of each other, is not known. Walter Wheldon, who lived in Park Grove, Bootle, about a mile away from Hornby Road, was a confectioner, born in York in 1883 and married to Hannah. Walter and James Alfred shared a great grandfather in Robert Wheldon, Walter's grandfather being John a younger brother of

James Wheldon Senior. Another Wheldon, Robert, also a confectioner, lived during the 1920s in Oban Road, Anfield. This was probably Walter's brother who had originally moved from York to Stockport and then possibly onto Liverpool, but again it is not known whether he knew of or communicated with the Wheldons in Hornby Road.

During 1924 Wheldon was in increasing discomfort as a result of the prostate cancer. He had two spells in the Royal Infirmary during the year, and was nursed at home by Madge, with the help of Nurse Tyson. His death in November of that year was marked by many obituaries in the scientific press, evidence of the high regard in which he was held.

Surviving among family papers are records, including some sketches, that Wheldon made of a family outing to the coast, and these give one or two insights into family life at that time; for example, they show him to be a pipe smoker and we are introduced to another member of the family, Jim the dog – once again the Wheldon family have not been very imaginative when choosing names! The exact date of the outing is not given but it is probably around 1912-15. Catherine is still alive so it places it before September 1915 and it is mentioned that Madge met the party at Bootle, so it was probably after 1912 since she seems to be an established member of the family. The party consists of Wheldon, Catherine, Doris and Madge. It is probable that Harold is already married and in Warwickshire and Bill may have been in Sierra Leone. The trip appears to be to the beach and sand dunes on the Lancashire coast between Liverpool and Southport. The sketches suggest that the dunes were near to the railway, so that places the destination between Hall Road and Birkdale, which includes Hightown, Formby, Freshfield and Ainsdale, all places regularly visited on botanical trips. Those destinations were (and are) on the Southport line and the nearest station on that line to Hornby Road was Marsh Lane & Strand Road (now Bootle New Strand). This is a walk of about a mile and a quarter and in those days would be partially over fields and footpaths. One very familiar landmark at the time would be the gas works in Marsh Lane, which appeared until recently very much as they did in 1912-15, although demolition work had started at the end of 2010. Many other buildings in the area, including Marsh Lane Station, suffered in the Blitz in May 1941.

One other point of interest is that the party meets up with two of Wheldon's botanical friends on the sand dunes, Dallman and Waterfall. Dallman, based on a known photograph, seems to be the thinner of the two gentlemen in the sketch. The other gentleman could be William Booth Waterfall, a fellow member of the Moss Exchange Club. Waterfall, who died in 1915, was from Bristol so must have been visiting the Liverpool area. However, it is more likely that it was Charles Waterfall FLS (1851-1938), a member of the LBS, who lived at Shavington Avenue, Chester; it appears that Dallman and Waterfall are on a botanical expedition. In the picture, Wheldon is directing them away from a sand hill behind which Doris and Madge are changing for the beach!

The day appears to start with Catherine setting off for Marsh Lane Station, ahead of Wheldon and Doris, who catch up with her later. These scenes emphasise the undeveloped nature of the area between Hornby Road and Bootle, very different from today. They walked down Bootle Style Lane, a footpath across fields roughly where Aintree Road is today, before joining the already developed Marsh Lane. On arrival at the station, Jim the dog caused a scene and here Madge joined the party. Why she joined here is not clear; she may have been living in Bootle at that time, but she returned with the party to Hornby Road and it is more likely that she was already living in Orrell Park. It is probable that she had come straight from work; she worked in the Cotton Exchange Buildings at that time, which was adjacent to Exchange Station, and it was a short journey from there to Marsh Lane, four stations down the line. At this time there was a very intensive service on the Southport line at about ten-minute intervals, so she could have been in Bootle very quickly. This suggests the trip may have been on a Saturday afternoon, as she may have come straight from work as many offices worked until midday on Saturdays in this period. If it was a Saturday, this might be an alternative explanation of the absence of the brothers, particularly of William, a keen cricketer, although if this was the case she might have preferred watching him play rather than going to the seaside with the family. As mentioned above, it is also possible that William was in Sierra Leone, in which case the trip would date after February 1914. The sketches also illustrate the fashions of the time: the straw boater is the standard headgear for men in their leisure time and the dresses (and bathing costumes) worn by Catherine and the two younger ladies are typical of that era.

Once on the train, Jim continued to make trouble and on arrival at their destination Doris and Madge change for the beach, interrupted by Dallman and Waterfall. A swim was followed by a picnic on the sandhills. Wheldon enjoys his pipe and the two girls smoke cigarettes. For those who knew Madge later in her life this seems extraordinary as she believed very strongly that women should not smoke! The pictures also suggest that Madge and Doris were close friends, although at times in later years there was some friction between them; however, they stayed in touch for the rest of their lives. After tea there was more swimming followed by a weary journey home up the hill past the gas works in Marsh Lane. The day ended with Wheldon sitting in the garden at Hornby Road with his pipe, a whisky and soda and the *Echo*, with Jim the dog at his feet.

It is likely that this sort of day out was a common event as lengthy family holidays would not have been a regular thing owing to Wheldon's botanical activities, although it is possible that some family holidays were combined with botanical expeditions, such as to Scotland in 1905 when Harold was certainly present, and a record exists (in *The Lancashire Naturalist* in 1909) of a trip to the Isle of Man when Wheldon, Catherine and Harold were present together with J. W. Hartley and Mrs Hartley (no mention was made of Doris and William being on the trip). Another record exists of Doris being on the island in 1917, presumably with other members of the family. The pictures refer to Wheldon as "Pa". The family always called him this, and in later life Madge, when recalling her father-in-law, would still use that title.

Wheldon's self-portraits show him to be of average build, bald and wearing a trimmed beard. Photographs of him confirm this. He lost his hair at an early age (as did his son William) and the photographs show him ageing mainly by the colour of his beard. Photographs also suggest that he had blue or light-coloured eyes.

Setting off for Marsh Lane

Scene at Marsh Lane Station

In the train

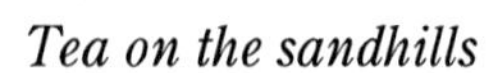
Tea on the sandhills

Dallman and waterfall appear

Homeward bound

Heading home past the gas works

Relaxing at Hornby Road

CHAPTER 5

THE NEXT GENERATION

James Alfred Wheldon and Catherine had three children. Their eldest was Harold James, born in York on 7 July 1888. He lived with his parents at 60 Hornby Road until 1912 when he married Mildred Ethel Dunning on 9 April of that year at the Brooke Road Church. After their marriage, they moved south to Warwickshire, firstly to Kenilworth and then from 1914-20 to Ryton on Dunsmore. Thereafter Harold moved to the London area where he remained for the rest of his life, firstly in Barnes and then at Thames Ditton and Sunbury. Harold and Mildred had three sons, Basil James (b. 1913), Roland Alfred (b. 1915) and John Neville (b. 1920), and a daughter Mildred (b. 1916), who in 2009 was still alive in New Zealand. All the children were born in Warwickshire, although they all came back to the Liverpool area. Basil was married in Formby in 1937 and died in 1965 in Ormskirk, Roland was married in Fazakerley in 1946 and later lived in Yew Tree Road, not far from Hornby Road; he died in 1981. Mildred was married in Aughton in 1937 and later moved to New Zealand. Sadly, John did not survive the war.

Harold's marriage did not last and he later (1931) fathered a second daughter, Pamela Gwendoline, with Winifred Louise Canney. Subsequently he lived with Gertrude Dunston who changed her name to Wheldon and lived with Harold until his death in 1945. Gertrude had been a friend of Doris Wheldon during her time in China, and after Harold's death, Gertrude and Doris lived together until Doris's death in 1967.

Harold was something of a polymath. He shared his father's interests and, as an accomplished water colourist, he illustrated some of his father's work. It is possible that Wheldon's relationship with Harold was similar to that of his own father, the two sharing and enjoying similar interests. Harold also made a study of mosses when he was a young man and later, with the encouragement of his father, he started to make a study of fungi. Wheldon Senior had an interest in fungi, but didn't have time to study them fully due to his work in other areas of botany. In 1914, Harold published a "Key to the British Agaricineae" which ran to thirty pages and was illustrated by Harold himself. Previously he had contributed an article "Lancashire Ascomycetes" in the *Journal of Botany*. Further articles followed on the fungi of the Isle of Man, the Scottish Highlands and the Lancashire coast, and these were illustrated by Harold and his father. Harold was a member of the British Mycology Society and is recorded as attending their meeting in Yorkshire in September 1918. His father also hoped to attend this meeting, but could not.

Upon his death, James Alfred Wheldon had left an unfinished article on the "Lichens of the Isle of Man". Harold intended to complete this article, but in the event did not and it was completed by Wheldon's friend and colleague J. W. Hartley and published in 1927. Harold's interest in botany declined after the First World War and he developed his interest in painting in oils (he spent time in Devon painting) and he also had a boat on the Thames (called the Wagtail).

Professionally, Harold was an accountant, and author of several books on accounting, law and economics. He wrote six books for the Bennett College, Sheffield from

1924 until 1935, and several titles published by McDonald and Evans during the 1930s and 1940s, some of which became standard works until quite recent times. Harold was a chartered secretary, a fellow of the Institute of Cost and Works Accountants, an associate of the London Association of Accountants, and later a fellow of the Royal Economics Society. He also gained a B.Com from the University of London. Whilst working as an accountant, he was also employed during the 1930s as a lecturer in costing at the City of London College. During both Wars he held official positions: in the First World War he was an auditor to the Ministry of Munitions and in the Second World War he worked as an accountant for the War Office. In his work as an accountant he was from time to time involved in the disposal of stock from bankrupt businesses. Jim remembers that Harold had asked Bill to store a quantity of radios and clocks which Harold had acquired from a bankrupt supplier. Harold also seemed to have some connections with Italy. It is possible that he had an office there. In addition to his work as an accountant and his interest in botany, Harold was an expert on contract bridge. He wrote a book *Complete Contract Bridge Simply Told* which was published in 1931 and ran to three editions. The book was published by Castell Brothers Ltd of London (part of the House of Castell, suppliers of bridge cards and accessories), who were based at the Pepys works in Clerkenwell.

Apparently Harold did not join the forces in the First World War, although he would have been of the right age. One possible reason may have been his health, and a possible heart condition. Jim remembers the last time Harold and Bill met. Bill was then working in Chester at the Lead Works and Harold visited him there during the later years of the Second World War. Bill remembered watching Harold walking back to the nearby Chester General Railway Station and stopping every few yards to catch his breath. Alternatively, his work for the Ministry of Munitions may have meant he was not called up. Harold died at home on 28 October, 1945; he was sitting in the garden and when Gertrude called him in for lunch, he did not respond – he had died of a heart attack. His published works are listed in Appendix 2.

The Wheldon's second son was William Alfred, sometimes known as Bill, who was born in York on 4 September 1889. Unlike his brother, William does not appear to have had any interest in botany and was less academically inclined than either Harold or his sister Doris. He was initially a clerk working for a provision merchant and it was probably as a result of this employment that William went to Sierra Leone in 1914. He first emigrated to Sierra Leone on 18 February 1914 to take up employment as a buyer of palm oil. He remained there, apart from a couple of trips home, until just before his son Jim was born in April 1920. A family tradition is that when in West Africa he had a pet leopard. His time in Sierra Leone is best described from the existing records of his membership of the Rokell Lodge of Freemasons in Freetown, into which he was initiated in December 1914. He was in Sierra Leone for most of 1915 and 1916, including at the time of his mother's death, returning to Liverpool in October 1916; but he was back in Africa by March 1917. He returned again to England in May 1919 for his marriage to Madge in July 1919, returning after a honeymoon in Keswick to Sierra Leone for one last time before returning in March 1920, just prior to the birth of his son Jim. His courtship with Madge must have been strange one as for five years before their marriage he was only in Liverpool for a few months and she never went to Africa.

On returning to Liverpool he continued to work for his employers in the Royal Liver buildings. It is thought that he worked for a company called the British West Africa Company, but no record has been found of a company of that name in the Liver Buildings at that time, although there were several with West African interests. Declining an offer to return to Sierra Leone, he gained employment with Rowe Brothers, originally a firm of builders' merchants in Pall Mall, Liverpool. The firm later became lead manufacturers and established a lead works in Hawthorne Road, Bootle where Bill worked. This establishment later became a separate company, the Librex Lead Co Ltd., by which time both it and Rowe Brothers, and another Liverpool firm of paint manufacturers, Goodlass Wall (for whom Bill may have worked at one time), were part of Associated Lead Manufacturers. Also part of ALM was the long-established firm of Walkers Parker who owned the Chester Lead Works. It was to Walkers Parker's plant in Chester that Bill was moved on the eve of the Second World War in 1939; he was employed there as a company secretary. Bill and Madge lived at 26 Marchfield Road, Orrell Park until 1925 when they moved "round the corner" to the then new 64 Stanley Gardens. Their daughter Joan was born here on Christmas Day 1926. In 1930 the family moved to 10 Cranmore Avenue, Crosby, in order to be closer to the Merchant Taylors' School where Jim had got a place. The family moved to Upton, Chester in 1939 and Bill died suddenly of a heart attack, aged fifty-seven, at their house at 46 Flag Lane on 24 January 1947. Like his brother, Bill seems to have avoided the First World War, probably because he was overseas but during the Second War he was a colour sergeant in the 71st Cheshire Home Guard, based at an anti-aircraft battery near to his home in Upton. He was fond of crown green bowls and for most of his life he was a keen amateur cricketer, opening both bowling and batting in local league cricket. Madge came to share his interest, learning to be a scorer. Madge died in Chester in 1980 and Jim passed away in November 2009 at the age of eighty-nine. Joan died as this account was being written in November 2010; she was eighty-three.

All four of James Alfred Wheldon's grandsons served in the Second World War. Sadly John did not survive the war as he was killed at RAF Kinloss in 1945, but the other three saw action in some of the most significant theatres of war. Basil was at Dunkirk, Roland was at Arnhem, and Jim was in HMS *Onslow* during the battle of the Barents Sea in 1942.

Wheldon's daughter Doris Mary was born in Ashton under Lyne on 18 July 1891. By 1911 she was working as a shorthand typist, but by 1919 or 1920 she had emigrated, first to Yokohama in Japan and then to Shanghai in China. Her responsibility for keeping house for her father passed to Madge after 1919, and Doris used this opportunity to travel. She was abroad at the time of her father's death, and at about the same time she was in hospital suffering from appendicitis. She made trips home in 1926 (when she stayed with Madge and Bill at 64 Stanley Gardens) and 1931, and came back to Britain in the late 1930s, possibly as a result of the Sino-Japanese War when Shanghai fell to the Japanese in 1937. She was initially employed as a shorthand typist, but she attained fairly senior commercial positions in Shanghai. On her return to Britain she settled in the London area and started a business, known as Mandarin, with her friend Gertrude Dunston. Mandarin imported silks and other fabrics from China, where they doubtless had built up many contacts. Doris and Gertrude were living with Harold at the time of his death at Roseland, Sunbury, Middlesex, and they continued to live there until Doris's death on 20 October 1967. Doris and Gertrude lived with Harold's daughter Pamela between Harold's death and her marriage.

Doris was an accomplished pianist, and at one time was a breeder of Havannah Rex rabbits. She never married, possibly because she was part of the generation of young women whose potential husbands were lost to the Great War. Until her death in 1967 she lived with Gertrude and kept in touch with her sister-in-law Madge through regular letters and the occasional phone call. It appears that Doris had some interest in her father's botanical work, and was a member of the LBS. In an article in the *Hong Kong Naturalist Magazine* in 1933, the eminent bryologist Hugh Neville Dixon (1861-1944) recounts that in 1922 Doris had collected a few mosses near Shanghai, and had sent samples home to her father. These had been passed on by Wheldon to Dixon who had commented upon them in the article "Mosses of Hong Kong and other Chinese mosses". Dixon and Wheldon were well known to each other, both being members of the Moss Exchange Club and Fellows of the Linnean Society (Dixon had proposed Wheldon for membership) and the two men corresponded regularly. Dixon had produced over 200 papers on mosses etc., so Wheldon would be very familiar with his work. Doris had earlier also sent some samples home to her father from Yokohama.

It is interesting to consider Wheldon's relationships with his children. He was obviously very close to Harold as they shared their interest in botany; Wheldon must have been very pleased and proud that Harold followed his example and published several articles and papers. Surviving letters suggest he was genuinely disappointed when Harold gave up his interest in botany. His relationship with Doris seems to have been a typical father/daughter one, probably enhanced by the difficult circumstances of Doris' birth. In letters to Arthur Dallman, his concern for Doris, while not explicit, is obvious. Wheldon would have seen little of William between 1914 and 1920, but after that he lived with him and Madge, and it would seem that William must have seen very little of his sister between 1914 and the late 1930s as his return from Sierra Leone coincided with Doris's departure for the far east.

CHAPTER 6

THE WHELDON HERBARIUM

When Wheldon died on 28 November 1924, he left an extensive collection of over 30,000 specimens, although not all of them were collected by Wheldon himself and it included specimens from many other collectors. Among these were several other members of the Wheldon family. Catherine and Doris both contributed lichens in 1912 (Lancashire) and 1917 (Isle of Man) respectively and there were numerous contributions from Harold (mainly bryophytes and lichens) from a variety of locations including South Devon, where he went to paint, and Warwickshire and Surrey where he later lived. There is also a record in The National Museum of Wales of a collection by Doris in Canada. It is possible that she visited Canada on her way to the Far East. It would appear that Wheldon's brothers George and Walter must have had some interest in botany as each had a contribution in the herbarium from 1881 and 1898 respectively. There were also several examples provided by a William Wheldon between 1850 and 1884 from locations in Yorkshire and Cumberland. This was probably the William who was born in 1803, a cousin of James Alfred's great grandfather Robert. These examples, together with others currently in the National Museum of Wales, date from before 1890 and thus seem to have survived the York fire.

Disposal of this herbarium was carried out by Harold from Barnes in south-west London, where he was then based. It is known that several institutions were interested in buying the collection, among them the British Museum, the National Museum of Wales and a museum in the Bronx, New York, which had asked for first refusal on the collection – further evidence that Wheldon's name was known beyond Britain. It appears that Wheldon had discussed the future of his herbarium before his death: he wished it to remain complete and in Britain, and had suggested to Bill and Madge that it was worth about £500.

The British Museum was interested in it, but did not want the whole collection, specifically the flowering plants, as it felt its collection of those was already extensive and since, as mentioned above, the family wanted the herbarium to remain complete, other potential buyers were sought. From shortly after Wheldon's death, the National Museum of Wales became interested in buying the collection and contact was made on 17 January 1925 by Mr H. A. Hyde, the keeper of botany. His initial enquiry was concerned with the size of the collection, to which Harold replied in early February 1925, stating that the collection consisted of the following numbers of specimens:

Flowering plants & ferns	15,000
Hepatics	2,000
Sphagna	2,700
Harpidia	1,600
Other mosses	7,900
Lichens	3,000.

He also stated that the collection included specimens from the Continent. Harold went on to write that "father was an expert of international repute in *Sphagna* and *Harpidia*" and that there was "not a finer collection in the world for extent and authenticity". He also referred in the letter to Professor McLean Thompson of Liverpool University who would confirm this view. Harold also stated in the letter that he had received about twelve obituaries of his late father from various sources. Professor Weiss of Manchester University later wrote to Hyde confirming the value of the collection and describing Wheldon as a very accurate "systematist".

Hyde's interest in Wheldon's herbarium had been alerted by Daniel Angell Jones, the secretary of the British Bryological Society, who lived in North Wales and had known and worked with Wheldon for many years, and indeed was probably one of the last to visit Wheldon before his death. He stated that as far as the *Sphagna* and *Harpidia* were concerned, there was no collection in the world to compare. He also said that Wheldon was in the first rank of lichenologists, an expert on flowering plants, and there was no collection of recent years to compare with Wheldon's. Further compliments came from Dr W. Watson of Taunton, who described Wheldon as the chief authority on *Sphagna* and *Harpidia* in England and a "careful determiner", and from Dr G. Claridge Druce of Oxford, a member of the BEC and at one time its treasurer, who confirmed that the collection would be of value.

During February 1925, arrangements were made for Hyde to visit 26 Marchfield Road to view the herbarium. The visit could only take place over a weekend as Harold was busy in London; it was during this time that he was studying for his B.Com at London University. It seems that Bill had no expertise or interest in the contents of the collection, so Harold would need to be present to deal with botanical and financial issues. Harold suggested that he travel up to Liverpool on the Saturday and return late on the Sunday, and that depending on train times, he and Hyde could meet at Crewe where Hyde would have to change trains from Cardiff. Harold also offered Hyde accommodation at Marchfield Road, subject, of course, to Madge's agreement. Later that month Harold stated that he wanted £400 for the herbarium. Hyde was surprised at that price and suggested that a going rate was nearer £300 and that some collections had sold for £1 per 100 pieces. On 21 February Harold wrote a rather strongly worded letter to Hyde saying that his father's collection was far superior to the "scrappy herbaria" which sold for the price Hyde had mentioned. He further stated that £400 was not excessive and hinted that he might look to sell abroad or split the collection, neither of which his father (or he) would have wanted.

It is not known whether the threat to split or sell abroad was serious or whether Harold, the accountant, was just trying to get the best price. However, it may also be that Harold was trying to conclude arrangements quickly by putting some pressure on Hyde. The collection was kept at 26 Marchfield Road and Bill and Madge were shortly to move from there to 64 Stanley Gardens, Orrell Park. If no sale were concluded before then, the herbarium would have to be moved to Harold in London, involving a double move which Harold wanted to avoid. Hyde's next letter on 23 February was apologetic and also wondered whether the price of £400 included the lichens. This enquiry stemmed from the fact that some further work was needed on the lichens and Harold, who retained an interest in them, would carry out this work. He agreed that they would be included in the £400.

On 4 March Hyde prepared a report to the Science Committee at the Museum, who would decide on whether to offer the £400. The report included some details of the material and the fact that he had sought opinions on the collection from Jones, Watson, Claridge Duce and others. He stressed the importance of keeping the collection together and in Britain, and pointed out that there was strong interest from the USA. He also valued it at £430 (doubtless to allow for some negotiation down to £400) and said that the money would go to the benefit of the Wheldon family.

By 12 March, Hyde had got the go-ahead and arrangements were made to transfer the herbarium to Cardiff. On 21 March, Bill wrote to Hyde confirming that he had sent eight cases by road transport (he thought that damage might occur if sent by rail) and that the lichens had been sent separately to Harold. Some correspondence arose in July 1925 about some missing mosses, but these eventually turned up. By the end of September 1925 the lichens were still with Harold and Hyde wrote to enquire whether he was still working on them or would he prefer to send them to Cardiff where the staff there could complete the work. On 12 November Harold agreed to this, and on the 28th (the anniversary of his father's death) he sent them to Cardiff via the Great Western Railway – obviously not sharing his brother's suspicion of railway transport. Interestingly in late 1925, Harold was working from the Pepys Works, Clerkenwell, the premises of Castell Bros who published his book on bridge six years later – it would be interesting to know of Harold's connection with this firm.

James Alfred Wheldon had recorded his collection in a book which Harold sent to Cardiff in March 1925. The book was returned to Harold later that month – its whereabouts today are unknown.

The collection still exists in Cardiff, and further collections made by Wheldon exist in several museums to this day, including Liverpool, Manchester and Bolton. It is interesting to note that the price paid for the herbarium, £400, would be around £17,000 today (based on the RPI) and that today it is unusual for herbaria to change hands for substantial amounts of money.

CHAPTER 7

THE DALLMAN LETTERS

During Wheldon's lifetime the post was the only real method of communication. The telephone system was in its infancy but there were several postal deliveries each day and people used letters and postcards to keep regularly in touch. Wheldon probably corresponded very regularly with family and botanical colleagues, so the volume of post would have been large. Fortunately some correspondence survives, notably a series of letters and postcards sent to Arthur Dallman between 1918 and Wheldon's death in 1924. It is significant that there are no letters before 1918 as after this date Dallman moved from his teaching posts on Merseyside to Manchester and then South Yorkshire, although apparently his wife did not move with him until about 1920 when they moved to Doncaster. It seems that they kept their house in Tranmere even after this move. All these letters shed some light on his domestic arrangements and family life, as well as on some of his botanical activities at the time.

A letter to Dallman in 1918 reveals that Wheldon was busy working on lichens collected in the Isle of Man (he had visited there in June of that year) and that Harold was at a meeting of the British Mycology Society in Yorkshire, an event Wheldon would like to have attended, but could not, showing that he still had an interest in fungi.

In April 1919 he wrote to Dallman bemoaning a lack of time and saying that he would have to cut back on his botanical correspondence and activities. The reason for this would appear to be pressure of work at the prison where it would seem that he was involved in some medical duties due to the lack of a deputy medical officer. He explained to Dallman that he was busy with blood tests (presumably of prisoners) and also in the administration of injections for syphilis to prisoners and for some unexplained reason to thirteen girls (presumably inmates). Wheldon is grateful that he has a deputy pharmacist to help him, but will be glad when a new deputy M.O. is appointed. It is possibly from this time that the story emerged that he sometimes, in that absence of a M.O. had to confirm death at an execution. No proof of this has been found, and it would seem unlikely. It is believed in the Wheldon family that at busy times Madge would help out in the pharmacy. In the same letter he tells Dallman of the financial problems of the MEC. It appears that the source of this was printing costs which Wheldon felt were excessive. He reports that the printers wanted £13 to print the annual MEC Report and he feared that they might have to reduce the size of the report or think about raising the subscription. He also mentions that he has had correspondence from Dr Tattershall and Professor Weiss about an unidentified task which he will not have time to take on. On a family note, the letter implies that Doris is living at home with her father and no doubt keeping house for him. This letter is written three months before the marriage of Madge and Bill, who then moved into 60 Hornby Road. This may have been the opportunity for Doris to have a life of her own, and she subsequently went to the Far East.

Wheldon wrote twice to Dallman in June 1921. He tells Dallman that he will be retiring from the prison service in a couple of months, and that as a result he will have to move from Hornby Road. He was therefore very busy looking for a house but so far had had no luck. He goes on to say that at current prices he may only be able to buy a small cottage and fears that as a result will have to dispose of the herbarium, which he reports contains about 30,000 specimens. In the letter he reveals that Mrs Britton of the Bronx Park Museum in New York had contacted him during the War and asked for first refusal on the cryptogams. He hoped that his new house would be large enough as he did not want to split his collection, and anyway in retirement he would have more time to devote to his Botany. He tells Dallman that Doris is now in Japan and that she has already sent some mosses and hepatics.

In the second letter of June 1921, Wheldon reports that he is still looking for a new house and that he has been told to move out by 31 August, although he hoped the authorities would let him hang on until October. He was still worried about what type of property he could afford. Apparently Dallman had offered to store his specimens for him, and he thanked him but hoped to keep his collection intact as he was looking forward to working on it during his retirement. This letter does not describe much botanical news, other than that the Rev. Gasking has had a stroke or seizure and that it would be some time before he returned to his botany, if he ever recovered. The letter also mentions the forthcoming publication of a paper on *Sphagna* in the *Journal of Botany*. This would be the "New British *Sphagna*" in July 1921. The letter reports that Doris is enjoying herself in Japan and that she had written several long letters (none of them of much interest to the Botanical Society!) about the Japanese culture and way of life and that she had sent him more samples and a butterfly! It would appear that at that stage Doris's time in the Far East was not expected to be a long one (although in the event she stayed there until the late 1930s) as he tells Dallman that he was grateful that Madge was looking after the house until Doris returned and that it was only because of that that Doris had been able to go. It would seem that the intention was that Doris eventually would return to look after her father. Madge and Bill (and baby Jim) were living at Hornby Road at this time.

Wheldon goes on in the letter to give news of his two sons. Harold seemed to have lost interest in botany which his father regretted; he attributes it to his war time activities as an auditor in the Ministry of Munitions, which involved him in "racing" around the country and having to store his books etc. for the duration. Harold was now "mad" on painting, especially landscapes in oils and has spent one or two holidays painting in Devon. In addition, he is now keen on boating, having moved near the Thames in Barnes, SW13. William (or Willie as he calls him in this letter) is in an office in the Liver Buildings working for the British West African Corporation. He had the opportunity to go back to Africa, but had declined due to his recent marriage and birth of Jim, despite the fact that he could make much more money there. Wheldon finishes the letter by saying that house hunting is much less fun that plant hunting!

The next record is a postcard posted from Rice Lane Post Office on 4 July 1922, in which Wheldon thanks Dallman for his congratulation on his MSc and expresses surprise at the award. He further said that he had been nervous about the ceremony, and was glad that it was now over. The only other news on the postcard was that Doris had moved from Yokohama and was now in Shanghai. By this time Wheldon would have been living at 26 Marchfield Road, Orrell Park, with Madge and Bill. It is not clear whether they or Wheldon owned the house. Madge always referred to it as their first home, although they lived at Hornby Road after their marriage, but soon after his death Madge and Bill moved to nearby Stanley Gardens, which suggests that the Marchfield Road property belonged to Wheldon Senior, or at least that he was renting it. On 8 August 1922 there was another letter to Dallman in which Wheldon thanked him for his kind invitation (to what is not known) but it seems that it was for a visit to Wales (Dallman was working on Floras of Flint and Denbigh), and goes on to explain that he was having a very busy time. He had some friends staying from Derbyshire and he had had a visit from his niece from Leicester who called in on her way to Morecambe (by car, which in the early 1920s would be worthy of note and also suggests that the niece was quite well off) with her husband. She was expected to call in again on her way back. In addition, Harold had been staying at Leasowe on the Wirral and Wheldon had been "running over there a good deal". It is not known whether this was a regular visit, but Jim always thought that Harold owned a property at Leasowe, but that has not been confirmed.

The identity of the niece is not given, and can only be speculated upon. Of Wheldon's siblings, George had at least three daughters, Mary, Helen and Bessie, who would have been twenty-nine, twenty-five and twenty-two years old at the time. Robert had a daughter Mary who would be 21, so any of these could have been married at that time. Any children of Lucie, Walter or Frederick would have been under twenty at the time of letter so it is unlikely that the visitor would have been any of these. It is also possible that the niece could have been from the Simpson side of the family. Septimus had had no children but Catherine's older brothers both had children of a suitable age in 1921. However, Thomas had seven children and Christopher had three, so it would be very difficult to identify the visiting niece.

Wheldon also says that due to his visitors he was unable to attend a gathering of the MEC at Dolgellau in August but he hoped soon to make a visit to the Isle of Man. (Which he did, in September, with Hartley.) Finally he invites the Dallmans to visit him at Marchfield Road. They obviously had not been there before as he gives directions to catch a train to Orrell Park Station, the nearest station to his new home. At this time the Dallmans would have been living in South Yorkshire.

None of the previous letters or cards give any indication that Wheldon's health was causing concern. But by 1924 he was suffering from prostate cancer and his final few letters to Dallman reflect this, and there is a noticeable decline in his handwriting. Interestingly, in a letter to Dallman in August, Wheldon mentions the fact that he had been advised to eat apples as a palliative for his condition, and he says that he felt better for doing so. Indeed even as he wrote the letter Madge was stewing some apples for him. On first reading this seems odd for a man with a background in science, but the old saying is "an apple a day keeps the doctor away" and even in the 21st century there are several websites available which claim the many health benefits of apples.

On 17 September 1924 he wrote to Dallman from Ward 9 in the Royal Infirmary. He felt he was improving and hoped to be home soon. He told Dallman that he had suffered a good deal and that the prognosis was not good as the cancer was malignant but he hoped that he had a couple more years. Doris would be home in about 18 months, he said, and he hoped to see her. He did not wish her to come home sooner as this would mean breaking her contract. Wheldon had enough medical knowledge to know that his situation was "hopeless" but hoped for a few more years. In fact he was dead in little more than two months.

There were two further letters to Dallman. The first of these on 23 October 1924 was actually written by Madge. She stated that Wheldon was in constant pain and found sitting in one position for any length of time difficult so he had asked her to deal with his correspondence. She said that her father-in-law, who had always been so active, found the enforced idleness very irksome, but that they lived in hope of some improvement. To add to the gloom, they had heard from Doris that she was in hospital in Shanghai suffering from appendicitis.

Wheldon's final letter to Dallman was written eleven days before he died and he managed to write it himself in a fairly shaky hand. The letter is quite distressing and moving and mentions his constant pain, relieved only by morphine (which destroys his appetite but he could not carry on without it), and that he feels himself to be a burden on those around him. He mentions that Albert Wilson has been in touch from his new home in the Conway Valley and that they have almost completed a paper on West Lancashire, a list of additions to the Flora published seventeen years previously, which will probably be offered to the *Lancashire and Cheshire Naturalist.* The letter gives news of Harold's latest accountancy publications and that he has been made a fellow of the "Society of Economics", and also that Doris has recovered from her operation in Shanghai. Wheldon says that he is expecting a visit from D. A. Jones and hopes that it is on one of his "good days" so that he doesn't have to have too much morphia which made him "muzzy".

The letters give the impression that the Wheldons and Dallmans were close friends, despite the age difference between them. Arthur Dallman was twenty years younger than Wheldon, and was only eight or nine years older than Harold and Bill. The later letters give fairly intimate details of Wheldon's condition and suffering, so the two must have been very good friends, although all the letters are addressed to Dear Mr Dallman or Dear Dallman, never Dear Arthur and are signed J. A. Wheldon but this formality would be the norm in those times.

Dallman's diaries still exist and he appears to have been a meticulous recorder. Some diaries contain his

Christmas card list (Mr and Mrs Wheldon included) and he made lists of all the letters he wrote. He appears to have been a regular visitor to Hornby Road, especially in the years before his marriage in the April of 1915 and there are mentions of 'supper with the Wheldons' and records of botanical trips with Wheldon, for example to Flintshire in July 1905, and as mentioned above in August 1906. Significant events are marked in the diaries. Catherine Wheldon's death is marked by a cutting from a newspaper, pasted into the diary "beloved wife of J. A. Wheldon. Service at Brook Road Methodist Church 1.30pm Wednesday 29th and at Walton Cemetery at 2pm" Such inserts in his diaries are rare, suggesting a particular affection and regard for Catherine. He attended the funeral. By 1924 Dallman was in Doncaster, but 28 November 1924 is marked in his diary, in stark contrast to his usual tiny writing, with the words 'Wheldon Dead' in large letters. No record exists of his attendance at the funeral (unlikely as he was in a teaching post in Yorkshire), but his record of his letters show that he wrote a letter of condolence to William on the 30th.

OBITUARIES

After his death on 28 November 1924, many obituaries of Wheldon appeared in the scientific and local press, Harold estimating that he had seen about a dozen, and it appears that he was genuinely held in high regard by his contemporaries. Professor McLean Thompson, in an obituary for the *Proceedings of the Linnean Society* wrote that the "loss to British Systematic Botany can hardly be estimated" and went on to write that "by his death the thinning ranks of British field botanists must suffer severely". *Nature* published a short obituary in December 1924 and referred to him as an "active and accomplished botanist" and a local newspaper described him as "an eminent Liverpool botanist". A major obituary published in the *Lancashire and Cheshire Naturalist* in February 1925 is certainly the most comprehensive: it starts with the statement that "science generally, but Lancashire botany in particular, sustains a great loss by the death of J. A. Wheldon, which took place on November 28th 1924". The writer of this obituary is not identified and it is interesting to speculate upon its author who evidently knew Wheldon well. The obvious candidate would be Dallman who was a prolific writer of obituaries, and connected with the *Lancashire and Cheshire Naturalist* but it is not listed in sources in the Dallman archive.

Most of the obituaries recognise Wheldon's reputation abroad in addition to Britain, and his willingness to help others with their botany. They also shed light on his personality and give an indication of the type of man that he was. McLean Thompson refers to his gentle disposition and deep sincerity, and also comments on Wheldon's resilience in adversity and how after the York fire he was undaunted by the tragedy and set himself to rebuilding his collection. A correspondent to the *Lancashire and Cheshire Naturalist* wrote of a "pleasant disposition and a congenial word for everyone" and that "he made fast friends with all who were privileged to go out with him". The same publication notes the following appreciation: "He was one of nature's gentlemen, and the work he has done for others in the cause of botanical knowledge is a sufficient testimony to his ability, great work, character and goodness of heart. His memory is revered and respected by all who had the pleasure of knowing him and his work".

An obituary was published by the British Bryological Society (previously the Moss Exchange Club) in 1925, written by Travis and Wilson in which their sadness at his death is apparent and they describe the loss to their society as "almost irreparable". Some additional facts emerge in this obituary not reported in others, for example that Wheldon spent a period of his professional training in Brighton. This was probably in the period between 1881 when he was in Scarborough and 1884 when he qualified in London. Given the close and long-standing friendship between Wilson, Travis and Wheldon there is no reason to doubt this. They also confirm Wheldon's abilities as a taxidermist and as an "excellent" ornithologist, and the influence of James Wheldon Junior who they note was a good botanist and ornithologist. Finally they say "of Mr Wheldon's personal character the writers of this notice can speak from long experience … he endeared himself to all who had the good fortune to make his acquaintance and although an exceedingly busy man, was always ready to help others".

To conclude, an obituary appeared in the local press on 1st December 1924, which was similar to, but more extensive than, the one published in the *Liverpool Daily Courier* that was mentioned in the introduction. It gives a very good summary of Wheldon's life.

DEATH OF MR J. A. WHELDON AN EMINENT LIVERPOOL BOTANIST

We announce with regret the death of Mr James Alfred Wheldon, M.Sc., A.L.S., I.S.M, which took place last week in his residence in Aintree. The late Mr Wheldon was for thirty years a pharmacist on the medical staff at HM Prison Walton, and at the time of his retirement three years ago was the senior pharmacist in the prison service. Shortly after his retirement he was decorated with the Imperial Service Medal.

Apart from his professional work he was an eminent botanist, widely known not only in British, but in international botanical circles. He was frequently consulted by botanists in many parts of Europe and North America, particularly in connection with those sections of cryptogamic botany in which he had long been an acknowledged authority, more especially mosses and lichens. He discovered and described many species new to science, and published many works and papers on the subjects he specialised in. A number of species have been named in his honour. Many of his publications have had special reference to Lancashire, and he was the joint author of the well-known "Flora of West Lancashire," a standard work.

HONOURED BY LEARNED SOCIETIES

Mr Wheldon was a foundation member of the Liverpool Botanical Society, and more than once its president. He was also a founder member of the British Bryological Society and for some years a member of the Liverpool Naturalists' Field Club. In 1901 he was elected as a fellow of the Linnean Society, and last year had the distinction of being elected an associate.

This is an honour for original botanical work and accorded only to a limited number of scientists. For his valuable botanical work he had conferred on him in 1922 the honorary degree of master of science by Liverpool University. On the occasion of the meeting of the British Association in Liverpool last year he served as vice president of Section K (botany) and contributed to the botanical chapters in the hand book "Merseyside", published in connection with that visit.

He loved to ramble on the fells and in the country lanes of Lancashire, and frequently tramped twenty to thirty miles a day, returning with heavy laden bags of specimens to which he devoted many hours with the microscope.

The internment takes place tomorrow (Tuesday) at 2pm at Walton Park Cemetery.

In Loving memory of
Catherine
The beloved wife of
James Alfred Wheldon
Died 26th September 1915
Aged 50 years
Also of
Septimus Herbert Simpson
Brother of the above
Died 15th March 1915
Aged 50 years
Also of
James Alfred Wheldon
MSc ALS
Beloved husband of the above
Died 24th November 1924
Aged 62 years

Copy of the inscription on the gravestone in Walton Cemetery

APPENDICES

The appendices contain a list of the known published work of J. A. Wheldon and of his eldest son H. J. Wheldon. The main source of information, especially for J. A. Wheldon's work, is the obituary published in the *Lancashire and Cheshire Naturalist* in February 1925. This information has been checked as far as is possible and those articles actually seen or listed elsewhere are marked *. Other examples may come to light in the future.

Copies of most of H. J. Wheldon's books are held by the Wheldon family today but the following does not claim to be a complete list.

The chronology of J. A. Wheldon's work can in general terms be seen to reflect the events of his life. Very little is produced between 1890 and 1897. This may be because he was just developing his skills as an author or may be because these would be very busy times with the York fire, moving to Liverpool, a new job at Walton and a very young family. In addition, both his parents died in Liverpool during this period. Between 1900 and 1907 the majority of work concerns West Lancashire at a time when he would have been researching the *Flora* with Wilson. After its publication, Wheldon's horizons seem to broaden with more work on the Isle of Man and Scotland. Perhaps a decline in the volume of work might be expected around 1915, with the deaths of Septimus and Catherine, but this does not seem to be the case. However, there are fewer articles in the years immediately after the First World

War. This was at a time when Wheldon was especially busy at the prison, helping out in the absence of a deputy medical officer, for example. There is a relative shortage of writing in the years 1921-22 at the time of Wheldon's retirement. This may be because he was preparing for his move from Hornby Road, and was considering the future of his herbarium.

Between his retirement and death, Wheldon seems to have put his extra leisure time to good use with a long list of articles appearing in the last year of his life. In later years his interest in *Sphagna* is clear. Similarly it is clear that H. J. Wheldon lost interest in botany after 1918, and focussed his attention on his career in accountancy. At the time of J. A. Wheldon's death in 1924, it was thought that his final work on Isle of Man lichens would be completed by Harold, but it appears that did not happen and the work was completed by J. W. Hartley, and was published in the *North West Naturalist* in 1927. Wheldon also presented papers to the Liverpool Botanical Society, for example in October 1909 on "the flora of the Higher Grampians". In the same year, his presidential address to the Society was "The position of systematic botany in the present day".

In addition to his botanical work, J. A. Wheldon never lost interest in wildlife in general and ornithology in particular. He made regular contributions of short articles to the *Lancashire and Cheshire Naturalist*, including "Habits of the Blue Tit" in October 1914 and "South Lancashire Bird Notes" in July 1915. He observed birds both in the garden of 60 Hornby Road and on the Orrell Park allotments. A further article covered Tree Pipits, and in 1914 an article on Manx birds. In June of the same year he reported the finding of a gannet washed up on the Lancashire coast. Even amphibians caught his interest, and he wrote about natterjack toads at Ainsdale (see above)

APPENDIX 1

PUBLICATIONS BY JAMES ALFRED WHELDON

Title	**Date**	**Journal**
Wild Flowers of the Week (Series)	1886	Yorkshire Chronicle
York Catalogue of British Mosses	1888	?
On Colouration of Birds' Eggs	1889	Science Gossip
Spirals in Plants	1889	Science Gossip
Subspontaneous Mosses	1889	Science Gossip
Casual and Alien Plants	1889	Science Gossip
List of York Casual Plants	1890	York School Report
Variations in Erythraea	1897	Science Gossip
The Mosses of Cheshire	1898	Journal of Botany
The Mosses of South Lancashire	1898	Journal of Botany
The Mosses of South Lancashire	1899	Journal of Botany

* The Mosses of West Lancashire	1899	Journal of Botany
New Lancashire Mosses	1900	The Naturalist
Additions to the Flora of West Lancashire	1900	Journal of Botany
Mosses of the Mersey Province	1900	The Naturalist
Notes on the Flora of Over Wyresdale	1900	The Naturalist
* Elgin Mosses	1901	Journal of Botany
* Mosses of West Lancashire	1901	Journal of Botany
* Additions to the Flora of West Lancashire	1901	Journal of Botany
* The North of England Harpidia	1902	Journal
West Lancashire Plants	1902	Journal of Botany
Additional West Lancs Mosses & Hepatics	1902	Journal of Botany
Rubi of Liverpool	1902	Green's *Flora of Liverpool*
* "Kantia Submersa" in Britain	1903	Journal of Botany
Plants new to Cheshire	1903	The Naturalist
Botanical Notes from the Lancashire Coast	1903	The Naturalist
Mosses of the Southport District	1903	British Association Handbook

Pollard Willow Fauna	1904	The Naturalist
West Lancashire Lichens	1904	Journal of Botany
Nesting Habits of Rooks	1904	The Naturalist
"A Gemmiparous Pterigynandrum"	1905	Revue Bryologique
Additions to the Flora of West Lancashire	1905	Journal of Botany
Flora of West Lancashire	1907	Book
Hepaticeae of South Lancashire	1907	Trans. Liverpool Botanical Soc.
A New Variety of "Sagina Reuteri"	1908	Journal of Botany
* Comments on the "Herpidia adunca" of Sanio	1908	Revue Bryologique
Invernessshire Cryptogams	1908	Journal of Botany
A New Lichen from Lancashire	1909	Trans. Liverpool Botanical Soc.
Flora of the Manx Curraughs	1910	Lancashire Naturalist
On Some Additions to the Manx Sphagna	1910	Lancashire Naturalist
Lichens (Chapter)	1910	*Guide to the Natural History of the Isle of Wight*
Marratts Collection of British Mosses	1910	Journal of Botany

New Lancashire Cryptogams	1910	Lancashire Naturalist
*New Lancashire Lichens	1910	Lancashire Naturalist
Inverness and Banff Cryptogams	1910	Journal of Botany.
*Social Groups and Adaptive Characters in the Bryophtya (with H J Wheldon)	1911	Lancashire Naturalist
Review of Smith's "British Lichens"	1911	Lancashire Naturalist
Notes from Periodical Literature	1911	Lancashire Naturalist
On the Batrachian Ranunculi	1912	Lancashire Naturalist
*The Sand Dunes in April	1912	Lancashire Naturalist
Some Alien Plants of the Mersey Province	1912	Lancashire Naturalist
A New Variety of Parnassia Palustris	1912	Lancashire Naturalist and Journal of Botany
Hale Point in May	1913	Lancashire Naturalist
The Dry Dune Flora in June	1913	Lancashire Naturalist
Review of the Lichen Flora of S. California	1913	Lancashire Naturalist
"Festuca Robboelloides" in S. Lancashire	1913	Lancashire Naturalist
"Hypnumlycopodioides" in Wallasey Peat Deposit	1913	Lancashire Naturalist

The Disappearing Peat Moss	1913	Lancashire Naturalist
*The Oenothera of the S. Lancs Coast	1913	Lancashire Naturalist
“Agrostisretrofacta” in S. Lancs	1913	Lancashire Naturalist
West Lancashire extinctions	1913	Lancashire Naturalist
*What is Lemon Thyme?	1913	Lancashire Naturalist
A New Lancashire Lichen	1913	Lancashire Naturalist
“HypericumDesetangsii” in Lancs	1913	Lancashire Naturalist
*[1]“Parnassia palustris” var. Condensata	1913	Lancashire Naturalist
*[1]HelleborineViridifolia in Britain	1913	Lancashire Naturalist
The Lichens of Arran	1913	Lancashire Naturalist
Stone Mites in West Lancashire	1914	Lancashire Naturalist
*[3] The Manx Sand-dune Flora	1914	Lancashire Naturalist
Alpine Vegetation on Ben-y-Gloe	1914	Lancashire Naturalist
[4]The Lichens of Perthshire	1915	Journal of Botany
*A New British Acrocordia	1915	Lancashire Naturalist
*“Bidens Minima”	1915	Lancashire Naturalist
*[1] The Lichens of South Lancashire	1915	Journal Linnean Society

Review of "The Lichens of Alaska"	1915	Lancashire Naturalist
*On Fissidens	1916	Journal of Botany
*A Westmorland Pilophorus	1916	Lancashire Naturalist
At Formby	1917	Lancashire Naturalist
*A Synopsis of European Sphagna	1917	Moss Exchange Club
*On the Collection, Taxonomy and Oercology of the Sphagna	1917	Lancs & Cheshire Naturalist
additions to the British Sphagna list	1917	Moss Exchange Club
*Further notes on the Manx Flora	1918	Lancs & Cheshire Naturalist
"Hypnumaduncumvar Wheldoni"	1918	Journal of Botany
Notes on Braithwaite's "Sphagnaceae Exsiccatae"	1919	Journal of Botany
Some Llanberis Lichens	1919	Journal of Botany
A Key to the Harpidioid Hypna	1920	The Naturalist
New British Sphagna	1921	Journal of Botany
Irish Sphagna	1922	Irish Naturalist
*Canon Lett's "Irish Sphagna"	1923	Irish Naturalist

Botanising in the Isle of Man (Lichens)	1924	Lancs & Cheshire Naturalist
Vegetation of S. Lancs Peat Mosses	1924	British Assn. Handbook
Vegetation of S. Lancs Sand Dunes Handbook	1924	British Assn.
New forms of Sphagna	1924	British Bryological Report
Notes on the Brya of the District	1924	Lancs & Cheshire Naturalist
"Acrocladiumcuspidatum" and its Varieties	1924	Lancs & Cheshire Naturalist
*Abnormal Plantains	1924	Lancs & Cheshire Naturalist
*Additions to the Scottish Sphagna 5	1924	Lancs & Cheshire Naturalist

[2] The Lichens of the Isle of Man (unfinished at time of death.)

[1] With W. G. Travis

[2] Completed by 1927 by J. W. Hartley

[3] With J. W. Hartley

[4] With A. Wilson.

5 Also in *The Journal of Botany* November 1924

APPENDIX 2

PUBLICATIONS BY HAROLD J. WHELDON

Botanical Works

Title	**Date**	**Journal**
Contribution to the Manx Fungus Flora	1908	Lancashire Naturalist
Some Highland Fungi	1909	Journal of Botany
Lancashire Ascomycetes	1911	Journal of Botany
Key to the British Agaricineae	1914	Book
Fungi of the Sand Dune Formation of the Lancashire Coast	1914	N/A
*Fungus Flora of Lancashire With J A Wheldon – (H J Wheldon illustrated)	1918	Lancs & Cheshire Naturalist
Social groups and adaptive Characters in the Bryophyta	1911	Lancashire Naturalist

A New Variety of "Parnassia Palustris"	1912	Journal of Botany and Lancs Naturalist

Accountancy/Economics/Law

* Bookkeeping and Accountancy Vol. 1	1925	Bennett College
* Bookkeeping and Accountancy Vol. 2	1925	Bennett College
* Auditing	1926	Bennett College
*Costing Systems and Accounts	1928	Bennett College
* Company Law	?	BennettCollege
* Economics	1935	Bennett College

Publication dates approximate. No date is given on the title page but some have been inscribed "to my brother William" by HJW and dated.

* Cost Accounting and Costing Methods	1932	McDonald & Evans
* Costing Simplified	1938	McDonald & Evans
* Applied Costing	1944	McDonald & Evans
Business Statistics and Statistical Method	?	McDonald & Evans

Bridge

*Complete Contract Bridge Simply Told	1931	Castell Bros. Ltd

APPENDIX 3

James Alfred Wheldon was a member of the Botanical Exchange Club for most of his life. He is recorded as having made around 200 collections during the years between 1876 and 1924, and an analysis of these collections can give some indication of his botanical interests during this period.

The first recorded collection by Wheldon was in Romanby, North Yorkshire, in 1876 when he was only fourteen or fifteen years old and it is evident that his interest in botany started at a very young age. In contrast, the final contribution to the BEC was in August 1924, three months before his death and at a time when he was quite seriously ill. The contribution was a collection made locally in Bootle.

A geographical analysis of collectors also gives an insight into Wheldon's activities over the years and largely confirms the comments made in Chapter 3 and Appendix One. Many of the collections were made locally in South Lancashire and Cheshire and cover most of the time that he lived in Liverpool. There were also a large number of collections made in West Lancashire and these were made mostly in the years between 1898 and 1907, and therefore coincide with the time that the *Flora of West Lancashire* was being prepared. One or two of these collections are also recorded as being made jointly, as might be expected, with Wilson. After the publication of the *Flora*, Wheldon's horizons broaden and after 1907 several visits were made to Scotland, especially the North East.

The first visit north of the border was actually made in July 1905 when collections were recorded in Perthshire. Interestingly one of these is credited to Harold, which suggests that possibly a family holiday was taken at that time. Other collections were made in Scotland in the summers of 1908, 1909, 1910 and 1911, apparently in the company of Wilson. September of 1908 also saw a visit to the Isle of Man.

Several collections were recorded in Wheldon's native Yorkshire in the period from 1876 until 1882, including in 1880 and 1881 at Scarborough during the time when Wheldon was living at Scarborough with Williamson. Interestingly two collections are recorded in the summer of 1891 in Yorkshire at a very busy time for the family, just after the birth of Doris and at the time of the move from Ashton to Liverpool.

Occasional visits were made to North Wales to Llandudno in the summer of 1894 and to various locations in Flintshire in 1906 and 1909. These visits could well have been with, or encouraged by, Dallman, who was particularly interested in the botany of North Wales, or with D. A. Jones. At least one visit was made to the Isle of Wight. A number of the BEC collections are recorded jointly. As might be expected, several are with Wilson. In addition to the 1905 collection with Harold, mentioned above, there are records of collections with A. A. Dallman in the summer of 1906 and with D. A. Jones in August 1902.

The table below shows the collections made by Wheldon in chronological order and classified by area, based upon Watson's vice-counties. A trend can be seen, showing the longer journeys made after the *Flora of West Lancashire* was published, and that local collections continued until his death.

Date	Location	Vice county
1876	Romanby	62
1878	Northallerton	62
1880	Scarborough	62
1881	Northallerton	62
1881	Scarborough	62
1881	Scarborough	62
1882	Pickering	62
1882	Pickering	62
1888	York	61
1889	Sandown IOW	10
1889	Sandown IOW	10
1889	Strensall Common	62
1891	Wallasey	58
1891	Scaleber	Wyork
1891	Exeter	3
1891	Ingleborough	64
1892	Southport	59
1892	Southport	59
1892	Southport	59
1892	Southport	59
1892	Wallasey	58
1892	Aintree	59
1892	Fazakerley	59
1893	Walton	59
1893	Walton	59
1894	Southport	59
1894	Hightown	59
1894	Walton	59

Date	Location	Vice county
1894	Great Orme	49
1894	Walton	59
1894	Southport	59
1894	Great Orme	49
1894	Fazakerley	59
1894	Eastham	58
1895	Walton	59
1895	Walton	59
1895	Walton	59
1895	Walton	59
1896	Walton	59
1896	Southport	59
1896	Aintree	59
1896	Ince Blundell	59
1896	Ingleton	64
1896	Eastham	58
1897	Ainsdale	59
1897	Ingleborough	64
1897	Liverpool	59
1897	Leasowe	58
1897	Birkdale	59
1897	Hightown	59
1897	Stonyhurst	59
1897	Northallerton	62
1897	Simmons Wood	60
1897	Simmons Wood	60
1897	Eastham Locks	58
1897	Eastham	58
1898	Hightown	59

Date	Location	Vice county
1898	Preston	60
1898	Southport	59
1898	Blackpool	60
1898	Aintree	59
1898	Aintree	59
1899	Bare	60
1899	Bare	60
1899	Bare	60
1899	Udale	60
1899	Halton	60
1899	Burton in Kendall	60
1899	Preesall	60
1899	Preesall	60
1899	Preesall	60
1899	Morecambe	60
1899	Morcambe	60
1899	Mitton	60
1899	Mitton	60
1899	Mitton	60
1899	Fleetwood	60
1899	Burton in Kendal	60
1899	Netherton	59
1899	Walton	59
1899	Walton	59
1899	Walton	59
1899	Aintree	59
1899	Netherton	59
1899	Thornton	59
1899	Morecambe	60

Date	Location	Vice county
1899	Freckleton	60
1899	Freckleton	60
1899	Preeshall	60
1899	Preeshall	60
1899	West Lancs	59
1899	Bootle	59
1900	Fleetwood	60
1900	Preston	60
1900	Aintree	59
1900	Quernmore	60
1900	Preesall	60
1900	Glasson Dock	60
1900	Fleetwood	60
1900	Fleetwood	60
1900	Cockerham Moss	60
1900	Bispham	60
1900	Lancaster	60
1900	Preesall	60
1900	Alston	60
1900	Fleetwood	60
1900	Fleetwood	60
1900	Blackpool	60
1900	Walton	59
1900	Glasson	60
1900	Glasson	60
1900	Longridge	60
1900	Grimsargh	60
1900	Grimsargh	60
1900	Preston	60

Date	Location	Vice county
1900	Winmarleigh	60
1900	Halton	60
1900	Preston	60
1900	Fazackerley	59
1900	Cockerham	60
1901	Fleetwood	60
1901	Aintree	59
1901	Bolton	60
1901	Aintree	59
1901	Aintree	59
1901	Aintree	59
1901	Fleetwood	60
1901	West Lancs	60
1901	Abbeystead	60
1902	Fleetwood	60
1902	Barnacre Moor	60
1902	Birkenhead	58
1902	Birkenhead	58
1902	South Lancs	59
1902	Knott End	60
1902	Birkenhead	58
1902	Birkenhead	58
1902	Birkenhead	58
1902	Birkenhead	58
1902	Cockerham	60
1902	Pilling	60
1902	Pilling	60
1902	Pilling	60
1902	Linacre	59

Date	Location	Vice county
1903	Aintree	59
1903	Aintree	59
1903	Rainford	59
1904	Rainford	59
1904	Rainford	59
1904	Rainford	59
1905	Killin	88
1905	Killin	88
1906	Llyn Helyg	51
1906	Llyn Helyg	51
1906	Llyn Helyg	51
1906	Ashton	60
1906	Heysham	60
1906	Heysham	60
1907	Aintree	59
1907	Aintree	59
1907	Fleetwood	60
1907	Walton	59
1907	Hatton	58
1907	Ainsdale	59
1907	Hightown	59
1907	Formby	59
1907	Walton	59
1908	Warton	60
1908	Freshfield	59
1908	Aviemore	96
1908	Ainsdale	59
1908	Walton	59
1908	Langness IOM	71

Date	Location	Vice county
1908	Churchtown	59
1908	Mossbridge	59
1908	Walton	59
1909	Aviemore	96
1909	Aviemore	96
1909	Churchtown	59
1909	Churchtown	59
1909	Gealcharn	96
1909	Gealcharn	96
1909	Banffshire	94
1909	Glen Eunach	96
1909	Glen Eunach	96
1909	Rufford	59
1909	Cwm	51
1909	Cwm	51
1909	Glen Eunach	96
1909	Cwm	51
1910	Bridge of Orchy	98
1910	Blackpool	60
1910	Ainsdale	59
1910	Cockerham	60
1910	Loch Rannoch	88
1910	Loch Rannoch	88
1910	Fairhaven	60
1910	West Lancs	60
1910	Cockerham	60
1911	Tyndrum	88
1911	Kirkby	59
1913	Walton	59

Date	Location	Vice county
1916	Walton	59
1921	Aintree	59
1922	Formby	59
1924	Bootle	59
1924	Birkdale	59

APPENDIX 4

Details of the Wheldon collections at the National Museum of Wales can be obtained from the publisher at 46 Flag Lane, Chester, CH2 1LE or can be downloaded from www.ypdbooks.com

Sources

Obituaries

Published by	Written by	Date
Lancashire and Cheshire Naturalist	N/A	Feb 1925
Proceedings of Linnean Society	J McLean Thompson	1925
The Pharmaceutical Journal	N/A	6 Dec 1924
The Chemist and Druggist	N/A	6 Dec 1924
Liverpool Daily Courier	N/A	1 Dec 1924
Nature Magazine	N/A	20 Dec 1924
British Bryological Society	Travis and Wilson	1925

Websites

herbariaunited.org

nature.com

southlancsflora.co.uk

Websites of British Bryological Society, Liverpool Botanical Society and Liverpool University

Other Sources

Proceedings of the Liverpool Botanical Society 1908-1922

The Lancashire Naturalist Vol II 1909-10 and Vol IV 1911-12

Dallman Archives at Liverpool Museum

Index of collectors in the Welsh National Herbarium by S.G.Harrison 1985

Archives of Liverpool Naturalists' Field Club at Liverpool Museum

Records of the Wheldon Herbarium at National Museum of Wales, Cardiff

Letters written by J A Wheldon held in the Liverpool Museum

Flora of West Lancashire written and published by Wheldon and Wilson 1907

Wheldon family archives including letters written by H.J.Wheldon